FRACTALS IN THE NATURAL SCIENCES

FRACTALS IN THE NATURAL SCIENCES

A Discussion Organized and Edited by
M. Fleischmann, F.R.S., D. J. Tildesley,
and R. C. Ball

FROM THE PROCEEDINGS OF
THE ROYAL SOCIETY OF LONDON

Published by Princeton University Press,
41 William Street, Princeton, New Jersey 08540
Copyright © 1989 by The Royal Society of London
and authors of the individual papers

Library of Congress Cataloging-in-Publication Data

Fractals in the natural sciences: a discussion/organized and edited by
M. Fleischmann, D. J. Tildesley, and R. C. Ball.
p. cm.
"From the Proceedings of the Royal Society of London."
ISBN 0-691-08561-7 (alk. paper)
ISBN 0-691-02438-3
(pbk.: alk. paper)
1. Fractals—Congresses. I. Fleischmann, M. II. Tildesley, D. J.
III. Ball, R. C. IV. Royal Society (Great Britain).
Proceedings of the Royal Society of London. Series A,
Mathematical and physical sciences.
QA614.86.F73   1990
514'.74—dc20   89-70015 CIP

ISBN 0-691-08561-7
ISBN 0-691-02438-3, pbk.

First Princeton Paperback printing, 1990

Reprinted by arrangement with The Royal Society of London.
These papers first appeared in *The Proceedings of The Royal
Society of London*, volume 423, pages 1-200, number 1864

Princeton University Press books are printed on acid-free paper,
and meet the guidelines for permanence and durability of the
Committee on Production Guidelines for Book Longevity of the
Council on Library Resources

10 9 8 7 6 5 4 3 2 1
10 9 8 7 6 5 4 3 2 1, pbk.

Printed in the United States of America
by Princeton University Press,
Princeton, New Jersey

FRACTALS IN THE NATURAL SCIENCES

A Discussion organized and edited by M. Fleischmann, F.R.S.,
D. J. Tildesley and R. C. Ball

(*Discussion held* 19 *and* 20 *October* 1988 – *Typescripts received* 28 *November* 1988)

[Three plates]

CONTENTS

	PAGE
B. B. Mandelbrot	
Fractal geometry what is it, and what does it do?	3
Discussion: A. Blumen	16
R. B. Stinchcombe	
Fractals, phase transitions and criticality	17
Discussion: E. Courtens	33
D. W. Schaefer, B. C. Bunker and J. P. Wilcoxon	
Fractals and phase separation	35
Discussion: J. S. Rowlinson, R. C. Ball, D. J. Tildesley	51
E. Courtens and R. Vacher	
Experiments on the structure and vibrations of fractal solids	55
M. Y. Lin, H. M. Lindsay, D. A. Weitz, R C. Ball, R. Klein and P. Meakin	
Universality of fractal aggregates as probed by light scattering	71
J. G. Rarity, R. N. Seabrook and R. J. G. Carr	
Light-scattering studies of aggregation	89
Discussion: D. A. Weitz	101
D. S. Broomhead and R. Jones	
Time-series analysis	103
R. C. Ball, M. J. Blunt and O. Rath Spivack	
Diffusion-controlled growth	123
Discussion: J. S. Rowlinson	132
P. Meakin and Susan Tolman	
Diffusion-limited aggregation	133
Discussion: A. Blumen	147
D. B. Hibbert and J. R. Melrose	
Electrodeposition in support: concentration gradients, an ohmic model and the genesis of branching fractals	149
Discussion: R. C. Ball	158

[1]

2   *Contents*

| | PAGE |

R. LENORMAND
Flow through porous media: limits of fractal patterns   159

P. PFEIFER, M. OBERT AND M. W. COLE
Fractal BET and FHH theories of adsorption: a comparative study   169

A. BLUMEN AND G. H. KÖHLER
Reactions in and on fractal media   189

Discussion: D. W. SCHAEFER   199

Fractal geometry: what is it, and what does it do?

By B. B. Mandelbrot

Physics Department, IBM T. J. Watson Research Center, Yorktown Heights, New York 10598. U.S.A.
Mathematics Department, Yale University, New Haven, Connecticut 06520, U.S.A.

[Plate 1]

Fractal geometry is a workable geometric middle ground between the excessive geometric order of Euclid and the geometric chaos of general mathematics. It is based on a form of symmetry that had previously been underused, namely invariance under contraction or dilation. Fractal geometry is conveniently viewed as a language that has proven its value by its uses. Its uses in art and pure mathematics, being without 'practical' application, can be said to be poetic. Its uses in various areas of the study of materials and of other areas of engineering are examples of practical prose. Its uses in physical theory, especially in conjunction with the basic equations of mathematical physics, combine poetry and high prose. Several of the problems that fractal geometry tackles involve old mysteries, some of them already known to primitive man, others mentioned in the Bible, and others familiar to every landscape artist.

FRACTALS PROVIDE A WORKABLE NEW MIDDLE GROUND BETWEEN THE EXCESSIVE GEOMETRIC ORDER OF EUCLID AND THE GEOMETRIC CHAOS OF ROUGHNESS AND FRAGMENTATION

Instead of attempting to introduce and link together the papers that follow in this Discussion Meeting, we prefer to ponder the question, 'What is fractal geometry?' We write primarily for the comparative novice, but have tried to include tidbits for the already informed reader.

Before we tackle what a *fractal* is, let us ponder what a fractal *is not*. Take a geometric shape and examine it in increasing detail. That is, take smaller and smaller portions near a point P, and allow every one to be dilated, that is, enlarged to some prescribed overall size. If our shape belongs to standard geometry, it is well known that the enlargements become increasingly smooth. Ultimately, nearly every connected shape is locally linear. One can say, for example, that 'a generic curve is attracted under dilations' to a straight line (thus defining the tangent at the point P). And 'a generic surface is attracted by dilation' to a plane (thus defining the tangent plane at the point P). More generally, one can say that nearly every standard shape's local structure converges under dilation to one of the small number of 'universal attractors'. The term 'attractor' is borrowed from dynamics and probability theory, and the even more grandiose term, 'universal', is borrowed from recent physics. An example of exception to this rule is when P is a double point of a curve; the curve near P is then attracted to two intersecting lines and has two tangents; but double points are few and far between in standard curves.

[3]

Standard geometry and calculus (which is intimately related to it) have long proven to be extraordinarily effective in the sciences. Yet there is no question that Nature *fails* to be locally linear. Indeed, the shapes of Nature are so varied as to deserve being called 'geometrically chaotic', unless proven otherwise. Unfortunately, 'complete' chaos could not conceivably lead to a science. This is perhaps why many of the oldest concerns of Man, such as concerns with the shapes of mountains, clouds and trees or, with the floods of the Nile, had not led to sciences comparable in effectiveness to the physics of smooth phenomena.

Though the term 'chaos' was not used, one can say that a second kind of chaos became known during the half century, 1875–1925, when mathematicians who were fleeing from concerns with Nature took cognizance of the fact that a geometry shape's roughness may conceivably *fail* to vanish as the examination becomes more searching. It may conceivably vary endlessly, up and down. The hold of standard geometry was so powerful, however, that the shapes constructed so that they do not reduce locally to straight lines were labelled 'monsters' and 'pathological'.

Between the extremes of linear geometric order and of geometric chaos ruled by 'pathologies', can there be a middle ground of 'organized' or 'orderly' geometric chaos? The author has conceived and outlined such a ground, and gave its study the name of 'fractal geometry', the fuller name being 'fractal geometry of nature and chaos'. It will be argued momentarily that fractal geometry is best viewed as a geometric language, new as of 1975, which incorporates as 'characters' several of the mathematical monsters of 1875–1925, and whose uses have now become so diverse, that it is possible to sort them out as poetry, strictly utilitarian prose and high prose.

FRACTALS ARE CHARACTERIZED BY SO-CALLED 'SYMMETRIES', WHICH ARE INVARIANCES UNDER DILATIONS AND/OR CONTRACTIONS

Broadly speaking, mathematical and natural fractals are shapes whose roughness and fragmentation *neither* tend to vanish, *nor* fluctuate up and down, but remain *essentially unchanged* as one zooms in continually and examination is refined. Hence, the structure of every piece holds the key to the whole structure.

An alternative term is 'self-similar', which has two meanings. One can understand 'similar' as a loose everyday synonym of 'analogous'. But there is also the strict textbook sense of 'contracting similarity'. It expresses that each part is a linear geometric reduction of the whole, with the same reduction ratios in all directions. Figure 1 illustrates a standard strictly self-similar fractal, called Sierpiński gasket. In the early days, the resulting terminological ambiguity was acceptable to physicists, because early detailed studies did indeed concentrate on strictly self-similar shapes. However, more recent developments have extended, in particular, to include self-affine shapes, in which the reductions are still linear but the reduction ratios in different directions are different. For example, a relief is nearly self-affine, in the sense that to go from a large piece to a small piece one must contract the horizontal and vertical coordinates in different ratios.

Can the process of taking parts be inverted, by replacing zooming in by zooming

FIGURE 1. The Sierpiński gasket: construction and self-similarity properties, static and dynamic. The four small diagrams show the point of departure of the construction, then its first three stages, while the large diagram shows an advanced stage. The basic step of the construction is to divide a given (black) triangle into four sub triangles, and then erase (whiten) the middle fourth. This step is first performed with a wholly black filled-in triangle of side 1, then with three remaining triangles of side $\frac{1}{2}$, etc. Perform a similarity (or more precisely a homothety), that is, an isotropic linear reduction whose ratio is $\frac{1}{2}$, and whose fixed point is either of the three apexes of the triangle that circumscribes the gasket. It is obvious by examining the large advanced stage picture, that each of the three reduced gaskets is simply one third of the overall shape. For this reason, the fractal gasket is said to have three 'self-similarity' properties. As defined, these self-similarity properties seem 'static' and 'after-the-fact.' However, it is also possible to reinterpret them as forming together a generalized dynamical system. In that case, the gasket becomes redefined as a dynamic 'attractor'.

away from the object of interest? Indeed, many fractals, including the strictly self-similar ones, can be extrapolated to become unbounded, and they can remain unchanged when the above process is reversed and the shaped is examined at increasingly rough scales. To gain an idea of the appearance of an extrapolated Sierpiński gasket, it suffices to focus on a very small piece of figure 1, and then to imagine that this piece has been blown up to letter size, and therefore the whole gasket has become so large that its edges are beyond visibility. This illustration shows that each choice of a small piece to focus upon yields a different extrapolate.

When the Sierpiński gasket is constructed as in figure 1, that is, by deleting middle triangles, one sees it has three properties of contracting self-similarity, as pointed out in the caption of figure 1. These properties appear, so to speak, as 'static' and 'after-the-fact', but this is a completely misleading impression. Its prevalence and its being viewed as a flaw came to us as a surprise. Therefore, it is good to stop and show how, knowing the same symmetries, it is easy to reconstruct the gasket 'dynamically' by a stochastic interpretation of a scheme due to J. Hutchinson. The basic principle of this scheme first arose long ago, in the work of Poincaré and Klein, and corresponding illustrations using randomization are found in our book (Mandelbrot 1982). Start with an 'initiator' that is an arbitrary bounded set, for example is a point P_0. Denote the three similarities of the gasket by S_0, S_1 and S_2, and denote by $k(m)$ a random sequence of the digits 0, 1 and 2. Then define an 'orbit' as made of the points $P_1 = S_{k(1)}(P_0)$, $P_2 = S_{k(2)}(P_1)$ and more generally $P_j = S_{k(j)}(P_{j-1})$. One finds that this orbit is 'attracted' to the gasket, and that after a few stages it describes its shape very well.

Surprise: simple rules can generate rich structure

How did fractals come to play their role of 'extracting order out of chaos?' To understand, one must go beyond simple shapes like the gasket or like the other fractals-to-be that mathematicians have first introduced as counter-examples. In these 'old' shapes, indeed, what one gets out follows easily from what has been knowingly put in. The key to fractal geometry's effectiveness resides in a very surprising discovery the author has made, largely thanks to computer graphics.

The algorithms that generate the other fractals are typically so extraordinarily short, as to look positively dumb. This means they must be called 'simple'. Their fractal outputs, to the contrary, often appear to involve structures of great richness.

A priori, one would have expected that the construction of complex shapes would necessitate complex rules. Thus, fractal geometry can be the study of geometric shapes that may seem chaotic, but are in fact perfectly orderly.

Let us, for the sake of contrast, comment on the examples of a related match between mathematics and the computer that arise in areas such as the study of water eddies and wakes. In these examples, the input in terms of reasoning or of number of lines of program is extremely complicated, perhaps more complicated even than the output. Therefore, one may argue that, overall, total complication does not increase in those examples, merely changes over from being purely conceptual to being partly visual. This change-over is very important and very interesting, but fractal geometry gives us something very different.

What is the special feature that makes fractal geometry perform in such unusual manner? The answer is very simple; the algorithm involves 'loops'. That is, the basic instructions are simple, and their effects can be followed easily. But let these simple instructions be followed repeatedly. Unless one deals with the simplest old fractals (Cantor set or Sierpiński gasket), the process of iteration effectively builds up an increasingly complicated transform, whose effects the mind can follow less and less easily. Eventually, one reaches something that is 'qualitatively' different from the original building block. One can say that the situation is a fulfilment of what is general is nothing but a dream: the hope of describing and explaining 'chaotic' nature as the cumulation of many simple steps.

FRACTAL GEOMETRY VIEWED AS A LANGUAGE

'Mathematics *is* a language.' (Josiah Willard Gibbs (speaking at a Yale faculty meeting ... on elective course requirements).)

'Philosophy is written in this vast book – I mean the Universe – which stands forever open to our gaze, but cannot be read until we have learnt the language and become familiar with the characters in which it is written. It is written in the language of mathematics, and its characters are triangles, circles and other geometrical figures, without which it is humanly impossible to understand a single word of it; without which one wanders in vain through a dark labyrinth.' (Galileo Galilei: *Il Saggiatore (The Assayer)* 1623).

'The language of mathematics reveals itself unreasonably effective in the natural sciences..., a wonderful gift which we neither understand nor deserve. We should be grateful for it and hope that it will remain valid in future research and that it will extend, for better or for worse, to our pleasure even though perhaps also to our bafflement, to wide branches of learning.' (Eugene Wigner 1960).

Inspired by the above quotes. the best is to call fractal geometry a new *geometric language*, which is geared towards the study of diverse aspects of diverse objects, either mathematical or natural, that are not smooth, but rough and fragmented to the same degree at all scales. Its history is interesting and curious. As is clearly indicated by such terms as 'Cantor set', 'Peano curve', 'Sierpiński gasket', etc. ..., several fractal 'characters' date to 1875–1925. They count among the least complex (hence least beautiful), and they have seen previous uses in other languages that have nothing to do with fractal geometry.

However, as a language addressed to its new goals, fractal geometry was born with Mandelbrot (1975), the first edition of our book *Les objets fractals*.

There is a profound historical irony in the fact that these old 'characters' of the new geometry had been among the 'monsters' to which we have referred earlier. The general monster is *not* scaling, and the fact that some monsters are scaling had not been singled out, because it was viewed as a 'special' property, therefore one that is not very 'interesting'. In the mathematical culture of the century that ran from 1875 to 1975, special properties did not warrant investigation.

Clearly, the study of order in geometric chaos could not arise as a specialized object of study, and a term to denote it did not become indispensable, until we had performed two tasks. (i) We saw that diverse rough patterns of nature – noise and turbulence and diverse geographical features – are geometrically scaling. Eventually we, and now many others, have identified geometric scaling in many other areas of nature, and then explored its consequences. (ii) We saw that the proper tool to tackle scaling in nature is suggested by some of the old 'monsters'. The old monsters themselves are not realistic models, but the construction of new fractals was immediately spurred by the new need, and more accurate models soon became available.

A language can be appreciated in diverse ways. For fractal geometry, the reasons can be sorted into five categories. artistic, mathematical, 'historical', practical and scientific. Let us discuss them in turn.

THE LANGUAGE OF FRACTAL GEOMETRY OFTEN
CREATES PLASTIC BEAUTY

We use the term 'geometry' in a very archaic sense, as involving concrete actual images. Lagrange and Laplace once boasted of the absence of any pictures in their works, and their lead was eventually followed almost universally. Fractal geometry is a reaction against the tide, and a first reason to appreciate fractal geometry, because the 'characters' it adds to the 'alphabet' Galileo had inherited from Euclid, often happen to be intrinsically attractive. Many have promptly been accepted as works of a new form of art. Some are 'representational', in fact are surprisingly realistic 'forgeries' of mountains, clouds or trees, while others are totally unreal and abstract. Yet all strike almost everyone in forceful, almost sensual, fashion. The artist, the child and the 'man in the street' never seem to have seen enough, and they had never expected to receive anything of this sort from mathematics. Neither had the mathematician expected his field to interact with art in this way.

In any event, fractal geometry shows that there is an unexpected parallel to the above classic quote from Wigner. We have been fortunate to witness the revelation of the "'unreasonable' and 'undeserved' effectiveness of mathematics as a source of enjoyable form."

FRACTAL GEOMETRY AND MATHEMATICS

One can also appreciate the language of fractals because it happens to have led to new mathematics that several groups of specialists find attractive. To the layman, fractal art tends to seem simply magical, but no mathematician can fail to try and understand its structure and its meaning. The remarkable aspects of recent events is that much of this mathematics, had its origin been hidden, could have passed as 'pure', in other words, as absolutely self-referential.

Many of the early pictures of fractals have mostly served to 'visualize' facts that had been obtained previously by abstract 'pure mathematical' thought. For example, one who examines the computer-generated pictures of Julia sets and Fatou domains, which are mathematical objects that concern the dynamics of 'rational iteration', cannot help being filled with deep humility and boundless admiration for the creative powers Pierre Fatou and Gaston Julia had exhibited in 1917-19. He may also be tempted to echo Georg Cantor's assertion 'that the essence of mathematics resides in its freedom'. Such arrogance has indeed characterized much of mathematics during the period 1925-75, but an examination of further examples of fractal art necessarily brings it to an end. To many mathematicians, the newly opened possibility of playing with pictures interactively, has turned out to reveal a new mine of purely mathematical questions and of conjectures, of isolated problems and of whole theories. To take an example, examination of the Mandelbrot set leads to many conjectures that were simple to state, but then proved very hard to crack. To mathematicians, their being difficult and slow to develop does not make them any less fascinating, because a host of intrinsically interesting 'side-results' have been obtained in their study.

Herein hangs a tale. Pure mathematics does exist as one of the remarkable activities of Man, it certainly is different in spirit from the art of creating pictures by numerical manipulation, and it has indeed proven that it can thrive in splendid isolation, at least over some brief periods. Nevertheless, the interaction between art, mathematics and fractals confirms what is suggested by almost all earlier experiences. Over the long haul mathematics gains by not attempting to destroy the 'organic' unity that appears to exist between seemingly disparate but equally worthy activities of Man, the abstract and the intuitive.

Nevertheless, fractal geometry is not (at least as of today) a branch of mathematics like, for example, the theory of measure and integration. It fails to have a clean definition and unified tools, and it fails to be more or less self contained. Calling it 'fractal geometry of nature and of chaos', immediately explains why it is more like the theory of probability. Both thrive only when they are not over-defined, when they do not mind using mismatched tools, and when they overlap heavily with many neighbouring endeavours. Moreover, the comparison with probability is not meant to imply a comparable level of development. While the fractalists have some reason for modest pride, their mathematical language keeps continuing to evolve and to expand with each new use. To find that a new use requires the improvement of some basic point is not an exception, rather an everyday experience.

FRACTALS AND SOME GREAT OLD 'PROBLEMS THAT POSE THEMSELVES'

There is a third reason to appreciate fractal geometry, whose validity is totally independent of the earlier two. This new reason involves the antiquity and the prevalence of the natural phenomena on which it has made a dent, either by descriptions that suffice to the engineer or by theories that satisfy the sophisticate.

The great Henri Poincaré drew, long ago, a distinction between those problems that a scientist chooses to pose, and those problems that pose themselves. (He then went on to praise Paul Painlevé for having identified a problem that nature has been attempting to pose, and for having tackled it.) To sort out all problems of Nature in this fashion is not an assignment that a prudent scientist will want to undertake. Besides, a scientist gains no merit by contemplating problems which his tools are too feeble to solve. Nevertheless, Poincaré's distinction is intuitively an essential one. Does a new scientific enterprise contrive new problems for the pleasure of solving them easily? Or does it, however short it may remain from achieving perfection, contribute to sharpening and understanding some problems that are very old

First example. To try and answer these questions in the case of fractal geometry, the best is, of course, to first allude to the mountains, the clouds, the water eddies and the trees. To the highest degree, the problems raised by their geometry deserve to be described as being 'problems that pose themselves'. Man-the-artist must have been pondering them forever, but Man-the-scientist did not know how to start tackling them. (At this point of the argument, an etymological digression comes to mind. The root of *geometry* is the Greek γεωμετρία. The customary translation, 'measure of the Earth', used to suggest to us that 'geometry' had once denoted the problems we now tackle by fractals. But we are no longer sure; did it denote a far-reaching 'measure of the Earth', or a practical 'measure of the land' (as claimed in Eric Partridge's *Origins*). The *Oxford English Dictionary* points out that 'in early quotations, geometry is chiefly regarded as a practical art of measuring and planning, and is mainly associated with architecture.')

Second example. We continue with quotes from the King James Version of the Bible.

> ... were all the fountains of the great deep broken up, and the windows of heaven were opened. And the rain was upon the earth forty days and forty nights. *Genesis* **6**, 11–12.
> ... there came seven years of great plenty throughout the land of Egypt. And there shall arise after them seven years of famine... *Genesis* **41**, 29–30.

Awe of the actual unpredictability of the weather obviously permeates the stories of Noah's Flood, and of Joseph son of Jacob, the interpreter of Pharaoh's dream of the Seven Fat and Seven Lean Cows. More down-to-earth records confirm that such occurrences are in fact common in meteorology. On the other hand, none of the standard tools of modelling can account for these symptoms. Arguing that this failure could not be 'fixed' by small changes led us, many years ago, to draw a certain sharp distinction, to coin for it a flippant but immediately useful terminology, and to introduce very different tools that eventually become integrated in fractal geometry.

When a natural phenomenon is such that its action is felt much of the time, yet most of its effects concentrate 'oligopolistically' in one or a few largest events, the phenomenon is now said to obey the *Noah Effect*. A phenomenon is said to obey the *Joseph Effect* when it involves 'trends' of arbitrary slowness, whose cumulative effect exceeds the effect of even the largest addend.

We cannot dwell upon our studies of these two effects, beyond saying that the tools we used in these studies were eventually redesigned to yield our fractal models of mountains and of clouds.

Third example. It is 'turbulence'. To announce that at long last 'a cure' for it has been found seems a way of securing publication and an audience, but there is no cure as yet, only small advances raising big new questions. Turbulence can be taken as a prototype of the fractal phenomena that are old and interesting, and at the same time resistant to analysis. Several of the decisive early steps towards fractal geometry, such as the introduction of multifractal measures, were taken in 1968–76 when we were working on turbulence. Multifractal and fractal tools have contributed their share to describing and understanding turbulence, but it has not been explained, yet.

Fourth example. It is far less widely known than the preceding three, but far more relevant to the topic of the present meeting. It concerns *fractal growth* models and *fractal aggregates*. The discovery of diffusion limited aggregation (DLA) by Witten and Sander has initiated one of the most surprising and challenging quests of the last ten years in condensed matter physics. A complete understanding of the fractal and multifractal properties of such aggregates remains an open and intriguing problem, to which we shall return momentarily. Yet, one who sees a fractal aggregate for the first time is likely to experience a feeling of déjà-vu. Figure 2, plate 1, suggests one reason why.

FIGURE 2. Viscous fingers in a chunk of amber. This photograph by Paul A. Zahle. Ph.D. is © 1977. National Geographic Society. It has appeared in the *National Geographic Magazine* in September 1977. pp. 434–435. To quote from the accompanying text by T. J. O'Neill. 'apparently a tiny crack developed in an already solidified lump of resin; fresh. sticky resin then began to fill the narrow crevice. Air also crept along the fracture plane. thus forming the array of mosslike pseudo-fossils. Some foreign substance – perhaps iron oxide – [provided color]' ... 'The classical Greeks called [amber] *electron*.' ... 'Not until when the Roman author Pliny made public his *Historia Naturalis*. was amber scientifically described as a product of the plant world.' Not until fractal geometry had become available. could the 'array of mosslike pseudo fossils' be recognized as an example of viscous fingering. and could become an object for quantitative study.

FRACTALS AND THE HARD-PRESSED ENGINEER

Yet another reason to appreciate the language of fractal geometry is in some way a restatement of the third. The new language promises to be effective in engineering. Self-styled sophisticates tend to either forget or spurn the needs of the practical man. One reason is that he does not have the luxury of waiting until the phenomena he chooses or is asked to try and control have been explained to the satisfaction of the sophisticates. Instead, he finds himself lost in Galileo's 'dark labyrinth', where he does not even know what signs to look for or what to measure.

The list of investigations where fractals matter already or promise to matter very soon from the viewpoint of phenomenology and of engineering is already long, but to comment in turn on each 'case' would be tedious and would generate little light. Let us, instead, give a few typical examples.

Rock-bottom, a prerequisite of engineering, and also a continuing goal of science, is to *describe* nature quantitatively. But everyone who has tried knows that *to see* is a skill one must learn, and that one must learn what to measure. All too many disciplines harbour the strong wish of becoming quantitative, but do not know even how to begin. One standard way is to ask new questions for known answers, that is to borrow procedures from disciplines that have already reached a quantitative stage, and to hold on to these procedures if they appear to be effective. One finds that procedures one could borrow are not particularly numerous. While the diversity of nature appears to be without bound, the number of techniques one can use to grasp nature is extremely small and increases very rarely. Therefore, the enthusiasm usually generated by the birth of a new technique and the desire to test it more widely is healthy, and must not be disparaged. In a number of notable cases, the new fractal additions to Galileo's geometric alphabet prove to be of great help in efforts to see and to measure. Let us examine a few cases.

Viscous flow through porous media, e.g. the flow of water pushing oil, has proven recently to admit to several régimes, one of which is a 'front', which is an effective and desirable configuration, and another one is 'fractal fingering', which is undesirable. After this range of possibilities had become known, a colleague of ours saw a very old article on viscous flow. Next to a photograph that could have been taken yesterday, a diagram meant to summarize what the original photographer had seen in his work. Unfortunately, what he had seen turned out to lead nowhere, but what he has smoothed away turned out to prove important, and it included the fractal features.

Our second example is from hydrology. There is no question that the design of aqueducts and of dams involves many aspects of the science of materials. But what about the variability of the water discharge in rivers? Even if a full climatological explanation were available for the long-run component of this very erratic process, tasks have been assigned to different professions in such a way that the availability, or lack of availability, of a climatological explanation cannot possibly matter to the water resources community. Yet, when, starting in 1963, we advanced a model of the long-run persistence in river discharges, we found that explanation was perceived as mattering very much. Our model having an infinite memory span. The first and most frequent questions were the following: 'Why choose so peculiar a model?' 'Has this model already been seasoned in the usual way, by being used in physics?' and 'What is the climatological explanation of this model?' In the water resources community, the most quantitatively inclined practitioners seemed intimidated by assertions of the primacy of explanation over everything else, afraid perhaps of hearing someone thunder 'But where is the science behind what you do?' We argued that it is best to perform each task in its time. The irony is, of course, that within a few years our model ceased altogether to be 'peculiar', because close counterparts were discovered, and soon adopted, in many chapters of 'mainstream' physics. These counterparts, as well as our model, have become building blocks of fractal geometry.

Our third example is from economics. To elaborate further upon the difference we see between the roles of fractals in engineering and in science, let us mention that security and commodity prices had been in the early sixties the topic of the first descriptive account that was to be later counted as fractal. It is a widespread assumption that price is a continuous and differentiable function of time. We claimed not only that it is not obvious and not only that it is contrary to the evidence, but that it is in fact contrary to what should be. The reason is that a competitive price should respond, in part to changes in anticipation, which can be subject to arbitrarily large discontinuities. (What happened on the Stock Exchanges on 19 October 1987 comes first to mind in 1988, but in the early 1960s such examples were viewed as things of the past.) On the basis of discontinuity combined with suitable self-similarity, we proposed a model of price variation, which had to incorporate the property of infinite variance of price increments. Our study of prices kept eliciting the comments already mentioned in the context of hydrology, the third one being phrased as 'Your models look fine, but how do you relate them to economic theory?' In moments of irritation, we are quoted are responding 'There is, as yet, no explanation for these findings; in fact *no* explanation could reasonably be expected to come from existing economic theory. After all, this theory has been growing for well over a century, and has yet to predict anything.'

The preceding sections show how difficult and unpopular is the task of defending the worth of scientific investigations that have not 'risen above' phenomenology. This difficulty is not new, in fact is very familiar in everyday life, where one can and must deal honestly with imperfection, while both preserving and keeping in check the dream of a More Heavenly City. In the context of fractal geometry, the Heavenly City of Explanation has already been reached in several cases, of which notable examples will be given. Everyone should rejoice, but it must immediately be acknowledged that other cases remain more like everyday life.

The scientific aspects of fractals are best discussed in two groups, those that are the most highly developed, and those which remain the most challenging.

THE ROLE OF FRACTALS IN THE THEORETICAL SCIENCES

The most highly developed examples of this role divide naturally into those which involve chance, and those which do not. The former are best exemplified by a chapter of statistical physics, called percolation theory, in which fractals have risen highest above rock-bottom phenomenology. In this chapter, the fractal description is admirably complete, and the physics have been shown to be ruled by geometry. by a small number of quantities, each of them the fractal dimension of some specified portion of a geometric shape called critical percolation cluster. Furthermore, the basic dimensions have been deduced from basic physics, and some even turn out to be rational numbers! The reduction of physics to geometry being one of the basic goals of physics, the role of fractals in percolation theory is close to perfection, even though the mathematician will interject that many basic facts that the physicist holds true have not been proven in full rigour, at this point. Most unfortunately, however, percolation theory is well outside the main highways of mathematical physics, so that it is best to move on.

A second source of completely understood random fractals is found in probability theory, namely in the sets of measures that illustrate limit theorems concerning the sums or products of random effects Brownian motion is the prime example. one that has preceded fractal geometry by decades but has been incorporated in its fold The Brown surfaces we have used as models of relief are another example. Self-avoiding random walk also shows every sign of being a fractal. though (again) the fact has not been proven in full rigour. The basic multifractal measures are obtained as limits of products of random factors.

Among non-random fractals of immediate relevance to physics, the best understood ones result from the fact that 'basins of atttraction' are usually bounded by fractal sets. Let us give an intuitive example. and then amplify this statement. Since a drop of water that falls on the United States and does not evaporate will eventually end in either the Atlantic or the Pacific, one can say that each of the two Oceans has a basin of attraction. The two basins are bounded by the Continental Divide, which our fractal model of relief happens to represent usefully by a fractal curve. More formally, consider a dynamical system that eventually converges to either of several limit states. Each limit state has a basin of attraction, and the typical situation is when the boundaries between the basins are fractal sets. The first example was advanced by Pierre Fatou. in the work that was already mentioned when we discussed the impact of fractals upon mathematics. In this case. which is why all the basins of attraction taken together are now called Fatou set. the boundaries of these basins being called Julia sets. The broad impact of the old Fatou and Julia work was felt only after it had been made part of fractal geometry. From the viewpoint of physics, however, the most important recent development may well be the proof that basin boundaries are also fractal in more general maps in continuous time.

Fractals and the great equations of mathematical physics

One can. last but not least, appreciate the language of fractals because it has already been used by master physicists and other scientists to produce beautiful works that add to our understanding of how the world is put together, and in particular of our understanding of the extent to which there is truth in the widely held notion that geometry rules physics.

What is generally perceived as the highest level of natural science is the study of the great old equations of mathematical physics. those of Euler Laplace.

Fourier and Navier–Stokes. The fact that these equations are *differential* implies a degree of smoothness that may seem *a priori* to forbid any connection with the rough and fragmented world of fractals. But it turns out that connections do exist, and that they are very important, even though some of them may have raised new questions in numbers exceeding the old questions they have already solved.

Our own contribution, which was the first interplay of fractals and differential equations, concerns the equations of fluid motion. We started with the old idea of Oseen, developed by J. Leray, that 'turbulence' is the name one gives to the effects of the singularities of Navier–Stokes equations. Granting this notion has allowed us to transform our fractal or multifractal models of turbulence into conjectures concerning the singularities of the equations. In specific mathematical terms, we introduced Hausdorff dimension as a new concern and a new tool in the study of the Navier–Stokes equations. The tool has proven attractive and the study to be fruitful, but by no means easy.

In the study of the Fourier equation, fractals have found a very attractive role in the notion of fractal fronts and of fractal foams of diffusion. The description we like best concerns a finite and discrete triangular lattice with a (hexagonal) ball placed on each lattice site. Let the abscissas of the balls run from $-x_{max}$ to $+x_{max}$, and start with the initial conditions where a ball is the colour of sand if its abscissa is negative, and is blue if the abscissa is positive. At each discrete instant of time, let 'couples' of neighbouring balls be chosen at random, independently of each other, and let their colours be interchanged. (The probability of being chosen is taken to be small, and if chance wants to choose a given ball as part of more than one 'couple', this ball can be left alone.) After a while, the relative proportion of blue balls will become a function of x, and the expected relative proportion is well known from the theory of diffusion of heat. Expectations, however, give an incomplete view of reality. What about the precise 'front' between the two colours? To define a front, reverse the colour in the sand islands entirely surrounded by blue sea, and in the blue lakes entirely surrounded by sand-coloured beach. This leaves us with a blue sea and a sand beach, separated by a shore line. Well, this shoreline happens to be a fractal curve, and its dimensions happens to be $D = \frac{7}{8}$. To be more precise, it does not becomes a fractal curve until the sites are down-scaled to infinitesimal sizes. As it wanders back and forth, this shoreline defines peninsulas that are attached to the beach by a single linking site. Imagine that these linking sites' colours are reversed. This will create a messy 'foam' separated from both beach and sea by wiggly curves: both wiggly boundaries happen to be fractal curves, and their dimension happens to be $D = \frac{4}{3}$. Higher-dimensional space diffusion creates an even more significant 'diffusion' foam. The proofs of the above assertions belong to the theory of percolation; hence, they are rigorous by the standards of physics, but continue to open purely mathematical problems.

The last but not least of the links between fractals and mainstream mathematical physics is raised by DLA and its many variants. It concerns the Laplace equations with interplay between, on the one hand, the solution of the equation at time t, and, on the other hand the displacement of the boundaries between times t and $t + 1$. The body of this book, however, contains so much about DLA, that it is hardly worth dwelling on it. This new problem promises to be difficult, and to stay with us for a long time. A conservative may even claim that it should have been mentioned in an earlier section concerned with phenomemology, but this is a minor issue. It is, we think, a feather in a cap of fractal geometry that it has created in advance the environment and the tools that had allowed the study of DLA to move forth as quickly as it had done.

By design, this paper is devoid of complete references. Also, all names of living persons were meant to be avoided; that is, the few that have escaped us are not meant in the least to disparage the persons who fail to be named. Any mention of our own work, again, merely demonstrates a regrettable lack of consistency. For references, see the papers in this Symposium, as well as the following list of books. It was roughly up to date in February 1989, but probably fails to be fully complete.

BIBLIOGRAPHY

Additional note An extensive, though far from complete, bibliography of fractals has been published as 'Resource Letter FR-1 Fractals', and a *Reprint book* has been announced by the American Association of Physics Teachers. The reference of the bibliography is Hurd A J. 1988 *Am. J Phys.* **56**, 969–975

Amann, A , Cederbaum, L & Gans, W. (eds) 1988 *Fractals. quasicrystals. chaos. knots and algebraic quantum mechanics* (Maratea, 1987 Proceedings). New York Plenum.
Avnir D (ed.) 1989 *The fractal approach to heterogeneous chemistry* New York Wiley
Barnsley, M F. 1988 *Fractals everywhere* Orlando, Florida Academic Press.
Barnsley, M. F (ed.) 1989 *Fractal approximation theory. J. Constructive Approximation* (special issue). New York Springer
Barnsley, M F. & Demko, S. (eds) 1987 *Chaotic dynamics and fractals* Orlando, Florida Academic Press
Cherbit. G (ed) 1987 *Fractals. dimensions non entières et applications* Paris Masson.
Dubuc, S 1987 *Atelier de géométrie fractale* (Montréal 1986 Proc.) *Annales des Sciences Mathématiques du Québec* (special issue).
Falconer, K J. 1984 *The geometry of fractal sets*. Cambridge University Press.
Family F. & Landau, D P (eds) 1984 *Kinetics of aggregation and gelation* (Athens. GA 1985 Proceedings) Amsterdam North-Holland.
Feder. J 1988 *Fractals* New York Plenum
Fischer. P & Smith. W. (eds) 1985 *Chaos. fractals and dynamics* New York M Dekker
Jullien. R & Bottet. R 1987 *Aggregation and fractal aggregation* Singapore World Scientific.
Jullien, R , Peliti, L . Rammal, R. & Boccara, N 1988 *Universalities in condensed matter* (Les Houches. 1988 Proceedings). New York Springer.
Lauwerier, H. 1987 *Fractals* (Dutch) Amsterdam Aramith
Mandelbrot. B. B. 1975. 1984 *Les objets fractals forme, hasard et dimension*. Paris Flammarion Also *Gli oggetti frattali* (Italian translation) Torino. Giulio Einaudi, 1987 Also *Los objetos fractales* (Spanish translation) Barcelona Tusquets. 1987
Mandelbrot. B. B. 1977 *Fractals form. chance and dimension* San Francisco. California W H Freeman
Mandelbrot, B. B. 1982 *The fractal geometry of nature*. New York W. H. Freeman. Also *Die fraktale Geometrie der Natur* (German translation) Basel Birkhauser 1987. Also *Fraktal kikagaku* (Japanese translation) Tokyo Nikkei Science. 1984.
Mandelbrot, B B and others (eds) 1984. 1985. 1986. 1987. 1988 *Fractal aspects of materials* (Extended Abstracts of Symposiums). Pittsburgh, Pennsylvania Mat Res. Soc
Mandelbrot. B B 1989 *Fractals and multifractals noise, turbulence and galaxies*, (*Selecta*. vol 1) New York Springer (In preparation)
Mandelbrot, B. B. & Scholz, C. H. (eds) 1989 *Fractals in geophysics* Boston Birkhauser *Special issue of Pure Appl. Geophys.* (In preparation)
Martin. J E & Hurd. A J 1986. 1987. 1988 *Fractals in materials science* (M R S Fall Meeting Course Notes) Pittsburgh Materials Research Society. (Out of print)

Mayer-Kress, G. (ed.) 1986 *Dimensions and entropies in chaotic systems* (Pecos River, 1985 Proceedings). New York: Springer.

Peitgen, H.-O. & Richter, P. H. 1986 *The beauty of fractals*. New York· Springer. Also *La Bellezza di Frattali* (Italian translation) Torino: Boringhieri 1988.

Peitgen, H.-O & Saupe, D. (eds) 1988 *The science of fractal images*. New York· Springer.

Pietronero, L. & Tosatti, E. (eds) 1986 *Fractals in physics* (Trieste, 1985 Proceedings). Amsterdam: North-Holland. Russian translation, 1989.

Pietronero, L. (ed.) 1989 *Fractals* (Erice, 1988, Proceedings). New York. Plenum.

Pike, E. R. & Lugiato, L. A. 1987 *Chaos, noise and fractals*. Bristol: Adam Hilger.

Pynn. R. & Skjeltorp, A. (eds) 1985 *Scaling phenomena in disordered systems* (Geilo, 1985 Proceedings). New York: Plenum.

Sapoval, B. 1989 *Les fractales*. Paris: Aditech.

Shlesinger. M. F., Mandelbrot. B. B. & Rubin. R. J. (eds) 1984 *Proc Gaithersburg Symp. Fractals in the Physical Sciences*. New York: Plenum, 1984. Special issue of *J. statist. Phys.*

Stanley, H. E. & Ostrowsky, N. (eds) 1986 *On Growth and form: fractal and non-fractal patterns in physics* (Cargèse, 1985 Proceedings). Boston and Dordrecht· Nijhoff-Kluwer.

Stanley, H. E. & Ostrowsky, N. (eds) 1988 *Fluctuations and pattern formation* (Cargèse, 1988 Proceedings). Boston and Dordrecht· Kluwer.

Stewart, I. 1983 *Les fractals* (Les Chroniques de Rose Polymath). Paris· Edouard Belin.

Takayasu, H. 1985 *Fractals* (Japanese) Tokyo Asakura Shoten. English translation: Manchester University Press (In preparation.)

Takayasu, H. M. 1988 *What is a fractal?* (Japanese) Tokyo. Diamond.

Tildesley, D. (ed.) 1989 *Fractals in the natural sciences*. This Symposium.

Ushiki, S. 1988 *The world of fractals· introduction to complex dynamical systems* (in Japanese). Tokyo: Nippon Hyoron Sha.

Vicsek, T. 1989 *Fractal growth phenomena*. Singapore. World Scientific Publishing.

Discussion

A. BLUMEN (*Universität Bayreuth, F.R.G.*). Could Professor Mandelbrot please comment on the fact he mentions in one of his papers on multifractals, that 'each $\rho(\alpha)$ has its own multiplicative factor M'? Is there any signature of $\rho(\alpha)$, say in a similar way as we obtain the central limit theorem for certain distributions and Lévy-forms for others?

B. B. MANDELBROT. For sums of independent identically distributed addends $\log M$, the large deviations distribution $\rho(\alpha)$, hence the function $f(\alpha)$, is fully determined by the distribution of the addends $\log M$, and conversely. Thus, in the strict sense, there is no universality whatsoever. To the best of my knowledge (but I do not know all that much), the same is true when the addends are dependent and non-identical. This fact about the multifractals is very significant in their theory, and must be recognized. I think that the usual formalistic approach had hidden it.

But what does it mean in practice? It need not necessarily mean much. In particular, the tails of the empirical $f(\alpha)$s depend on high moments, and are determined with very low precision. It is tempting to define, for any given $\epsilon > 0$, a 'class of approximate ϵ-universality' that would include all the M that yield the same $f(\alpha)$ 'within ϵ'. (The strip of uncertainty could be either vertical or horizontal.) Such a 'neighbourhood of $f_0(\alpha)$' would contain $f(\alpha)$s for which the slope at α_{max} and α_{min} is infinite. and other $f(\alpha)$s for which the slope is finite. The corresponding class of ϵ-universality for M may, or may not, contain random variables that are not particularly alike in other ways. The topic is wide open. It deserves to be examined carefully, and are right to be concerned.

Fractals, phase transitions and criticality

By R. B. STINCHCOMBE

Theoretical Physics Department, University of Oxford, 1 Keble Road, Oxford OX1 3NP, U.K.

Analogies between behaviour in fractal media and at phase transitions suggest deeper connections. These arise from the scale invariance of configurations at continuous phase transitions, making fractal view-points useful there and making scaling techniques necessary in both areas. These concepts, inter-relations, and techniques are illustrated with the percolation problem and the Ising spin system, which provide simple examples of geometrical and thermal transitions respectively. Diluted magnets involve both geometrical and thermal effects and, when prepared at concentrations near the percolation threshold, exhibit the influence of geometrical self-similarity on such varied processes as cooperative thermal behaviour and dynamics. Phase transitions in these systems are briefly discussed. The anomalous dynamical behaviour of self-similar systems is introduced, and illustrated by the critical dynamics of Heisenberg and Ising magnets diluted to the percolation threshold. These involve linear (spin wave), and nonlinear (activated) dynamical processes on the underlying fractal percolation structure, leading in the linear case to the magnetic analogue of 'fracton' dynamics and in the case of activated dynamics to a highly singular critical behaviour.

1. INTRODUCTION: FRACTALS, PHASE TRANSITIONS AND SCALE INVARIANCE

Typical fractals are characterized by structure on all scales of length (Mandelbrot 1977). Simple idealized examples much as the recursively constructed Cantor bar, Koch curve, or Sierpiński gasket (figure 1) go into (parts of) themselves under discrete scale changes, so having the property of 'discrete' self-similarity. Regular fractals with this property can occur naturally, for example in the spectra of incommensurate systems (Hofstadter 1976). However, real fractals are more often random and typically statistically self-similar. An example is the percolation network (figure 2), which will be discussed in §2.

The property of continuous, or discrete, or statistical, self-similarity is sufficient to cause scaling behaviour in these systems, which is characterized by scaling dimensions. The most basic of these is the fractal dimension d_f associated with the scaling of the mass. For fractals like the network shown in figure 2, d_f can be obtained in principle by drawing a succession of squares, of side length $L = L_1$, $L_2, \ldots$, measuring the mass $\Delta(L)$ of the network within each square, and noting that $\Delta(L)$ has the power law length scaling form

$$\Delta(L) \propto L^{d_f} \tag{1}$$

in which d_f provides the length scaling exponent, i.e. the effective dimensionality.

[17]

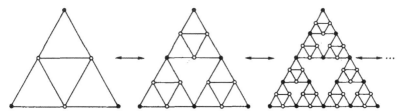

FIGURE 1. Recursive generation of Sierpiński gasket fractal.

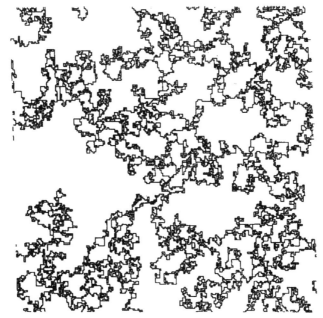

FIGURE 2. Backbone of percolation infinite cluster, for the case of
a bond-diluted square lattice.

A fractal dimension can similarly be associated with the regular fractal in figure 1. In this case its simple recursive generation allows us to consider an object that is not strictly self-similar, but behaves in a limited way as if it were. This object is obtained by running the recursive process only a finite number of times, until the smallest length occurring is a, say. Relations like (1) then only apply for $L_0 \gg L \gg a$, where L_0 is the system size, and the crossover to simple behaviour outside of this scaling window shows that the object is not strictly fractal.

Analogues to the features just described occur at continuous phase transitions. These encompass most of the common thermal transitions such as liquid–gas, superfluidity, superconductivity, magnetic and liquid crystal transitions (but not freezing) and the commonest geometrical ones, such as percolation. In these systems, the order parameter (magnetization, superfluid component, etc.) goes continuously to zero, and many physical properties show singular behaviour at the transition. An example is the variation of the order parameter itself, or the divergence of susceptibility or compressibility. The 'critical' behaviour is characterized by critical exponents describing the power law dependences of such properties on the distance (e.g. in temperature) from the transition.

The microscopic origin of the critical behaviour is the onset of long-range correlations: it is known from statistical mechanics, or from exactly solvable models, that a diverging correlation length ξ leads to singularities, and there is ample evidence of the diverging length from, for example, neutron scattering. The scale set by ξ for the microscopic configurations of the system thus goes away at the transition. The associated scale invariance is analogous to that in fractals, and the critical behaviour and exponents are analogous to the power law forms (see (1)) and associated dimensions characterizing fractals.

These analogies are deeper than so far indicated. The modern understanding of continuous phase transitions stems from the use of scaling methods, as introduced by Kadanoff, Wilson and Fisher (see, for example, Ma 1976). These techniques exploit the scale invariance, and the length scaling of parameters just as in (1). Examples will be given in §3. Moreover, the configuration of the system can be a genuine fractal at the phase transition, e.g. in geometric transitions like percolation, as illustrated in §2. The length scaling techniques have already transformed our understanding of critical phenomena, in particular allowing very complete descriptions of these cooperative effects, and explaining the identical critical behaviour of groups of systems. An example of such a group is the 'Universality' class including Ising spin systems, binary alloy, liquid–gas, etc., or that including XY magnet, superfluid, superconductor, certain liquid crystals etc. All the members of a universality class have the same critical exponents, as well as other universal quantities (amplitude ratios, scaling functions, etc.). Length scaling (Wilson 1972; Wilson & Fisher 1972) accounts for universality through the scaling-away of most parameters describing the system, leaving only a few 'relevant' parameters, whose scaling depends only on certain crucial aspects such as dimensionality of space and order parameter.

In the next section the percolation problem and Ising model will be given as simple examples of geometrical and thermal systems in which the scale invariance and 'fractal' or scaling characteristics can be identified and illustrated. From the remarks above, these examples provide the universal critical behaviour of large classes of systems.

2. Percolation and Ising systems

Percolation models were introduced (Broadbent & Hammersley 1957) in connection with percolation of fluid through porous media, and spread of disease through populations, e.g. an orchard. But the simplest percolation models occur in the context of dilute lattices. If sites of a regular lattice are removed at random, and occupied nearest-neighbour sites are regarded as connected, the size distribution of the resulting groups of connected sites (clusters) will depend on the concentration p of sites remaining on the lattice. In particular, if p is sufficiently large the largest cluster will span the lattice. This spanning cluster is called the infinite cluster in the infinite system limit. Its disappearance, as p is reduced through some critical value p_c, marks the percolation transition. At the transition the cluster distribution is scale-invariant, as can be demonstrated by comparing a typical configuration at p_c with a magnified portion of itself. The statistical scale invariance of such configurations implies that they are random fractals. The incipient infinite cluster at p_c, or its backbone (shown for a bond-diluted square lattice in figure 2), is also a random fractal.

Away from p_c, the cluster distribution is characterized by a 'correlation' length, ξ, which is typically the size of the largest finite clusters (see, for example, Essam 1980). This length diverges at the transition according to

$$\xi/a = C|p - p_c|^{-\nu}, \qquad (2)$$

(where a is the lattice spacing and C a constant), and the scale invariance is a consequence of the disappearance of the characteristic length. The critical exponent ν occurring in (2) is one of a family familiar in critical phenomena. Another is the exponent β for the order parameter, which, in site percolation, is the probability $P(p)$ that an arbitrarily chosen site lies on the infinite cluster (and so vanishes on one side of the transition):

$$P(p) \propto (p - p_c)^{\beta}. \qquad (3)$$

From (2) and (3), $(-\beta/\nu)$ can be regarded as the length-scaling exponent for $P(p)$. But $P(p)$ is the density of sites on the infinite cluster and so, by converting density to number via a volume factor, the fractal dimension of the infinite cluster is seen to be

$$d_f = d - \beta/\nu, \qquad (4)$$

where d is the euclidean dimension. So we obtain a relation between critical exponents and a dimension of a fractal occurring in percolation.

Similar arguments apply for the length-scaling dimensions at any continuous phase transition. So, for example, in the Ising magnet a correlation length ξ diverges at the transition according to an expression analogous to (2), and the specific magnetisation replaces $P(p)$ as order parameter in an equation like (3). The expressions now involve temperature in place of concentration, of course, and new values of the exponents β, ν. It follows that the length-scaling exponent for the total magnetic moment is $d - \beta/\nu$. One might think this is connected in some way with fractal dimensions of white (spin up) or black (spin down) regions in figure 3, which shows a typical configuration at the Ising transition. However, the white and black areas each scale with length like L^d (d is euclidean dimension). The total magnetic moment is proportional to the difference of the two areas. So $d - \beta/\nu$ is in this case best regarded as the length scaling of that property, rather than the dimension of some visualizable fractal. It should also be emphasized that length-scaling dimensions need not always be 'anomalous'. For example the length-scaling dimension of the (singular part of the) free energy in most thermal transitions is the euclidean dimension, and this leads to the property known as hyperscaling.

The correspondence of viewpoints we have tried to illustrate can be very fruitful

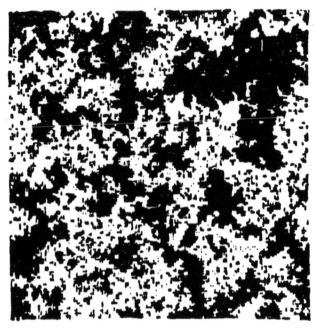

FIGURE 3. A typical configuration in a spin $\frac{1}{2}$ Ising model at its transition. Spin up and spin down regions are shown white and black respectively.

for both areas (fractals and phase transitions), for example in carrying across scaling techniques already developed for the other area (eg. renormalization group techniques as they originated for critical phenomena at phase transitions), or using geometric or fractal viewpoints at phase transitions. These possibilities are exploited throughout this paper and in many other recent studies (see, for example, Stanley 1985; Stinchcombe 1985 a, b).

3. SCALING

An example of a scaling technique that was first developed for phase transitions but that has been subsequently very useful for fractals is decimation. This was first introduced for thermal transitions (Kadanoff & Houghton 1975; Barber 1975) and shortly thereafter for geometric transitions, beginning with percolation (Young & Stinchcombe 1975). This technique will be used in the context of fractals in §5. A closely related scaling technique is blocking, which we illustrate below. This has its origins in Kadanoff's seminal work (Kadanoff 1966), but was not implemented until much later (Niemeyer & van Leeuwen 1974, thermal transitions; Harris *et al.* 1975, percolation). A simple version of blocking for percolation is as follows (Reynolds *et al.* 1977).

Figure 4a shows a site-diluted triangular lattice in which occupied or vacant sites are indicated by black or open circles respectively. Nearest-neighbour occupied sites are regarded as connected, as shown by the bonds in the figure. Each

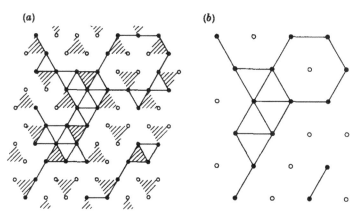

FIGURE 4. Scaling of a site percolation system by blocking: (a) original, and (b) scaled lattice. The sites of the scaled lattice correspond to blocks of three sites of the original lattice, and a 'majority rule' gives their occupation probability.

shaded block of three sites of the original lattice becomes a site of the rescaled lattice, figure 4b, and these 'rescaled' sites have been taken to be occupied if a majority (two or three) of the sites of the original trio is occupied. It can be seen that the connectivity of the original lattice has been maintained in the rescaled lattice, at least on a scale greater than the new lattice constant. Also that the rescaled (dilated) lattice is at a different concentration (p', say) from the original (p). Because p' is the occupation probability of a rescaled site, and this, by the 'majority rule', is the probability that two or three sites of the block are occupied, we have

$$p' = 3p^2(1-p) + p^3. \qquad (5)$$

This renormalization group transformation gives the rescaling of parameters accompanying a dilation of the lattice constant a by a factor $b = \sqrt{3}$. It provides, as follows (Young & Stinchcombe 1975) the critical condition (p_c) and exponent (ν).

At the transition ($p = p_c$) the system is scale-invariant so no parameter change is produced by the dilation. So p_c is that concentration which transforms into itself, i.e. a 'fixed point' of the transition. That yields $p_c = \frac{1}{2}$, which is the exact result for the lattice considered.

Near the transition the characteristic length ξ diverges according to (2). Denoting variables for the scaled system by primes, the corresponding equation for the scaled system has p' as given by (5), $a' = ba$, and $\xi' = \xi$ (because gross connectivity has been maintained on an absolute scale). Hence

$$b = \lambda^{\nu}, \qquad (6)$$

where $1/b$ is the inverse rescale factor for the left-hand side of (2) and the 'eigenvalue' λ ($= \mathrm{d}p'/\mathrm{d}p$ at p_c) is the rescale factor of $p - p_c$. Because both of these are known, (6) determines the critical exponent ν. The resulting value is fortuitously close to the exact value in this example. These methods can be made quite accurate by various refinements (Young & Stinchcombe 1975) or can be exact (e.g. the decimation example in §5). One of their great virtues is, however, the way they can capture the essence of a process or system in a very simple way. They have, of course, very wide applications and great generality, from the thermal contexts in which they originated to the fractal ones to be illustrated later.

4. STATIC BEHAVIOUR OF DILUTE MAGNETS; PROCESSES ON A FRACTAL

A system that involves all the ideas so far emphasized is the dilute magnet. It combines magnetic (thermal) aspects with geometrical ones (site percolation, via the dilution) and when diluted to the percolation threshold p_c it involves processes on a fractal (the percolation system at p_c). It is very rich in the variety of both static and dynamic scaling phenomena that can arise. This section provides a very brief account of some static aspects; dynamic phenomena are discussed at greater length in §§5 and 6.

Scaling techniques, like those referred to in the previous section, give the phase boundary (i.e. the transition temperature as a function $T_c(p)$ of the concentration, as shown in figure 5 for an Ising system) and the critical behaviour of a wide variety of static properties, such as correlation length, specific heat, magnetization, susceptibility etc. A review of such studies is given in Stinchcombe (1983a) and this also provides an impression of the typically excellent agreement with experiment on the many ideal real examples of diluted two- and three-dimensional Ising, Heisenberg and other spin systems. Instead of attempting a

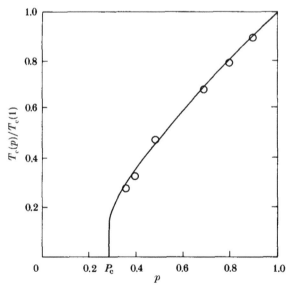

FIGURE 5. Phase diagram of a dilute Ising magnet a theoretical result for the dependence of critical temperature on concentration p is shown for comparison with experimental points.

comprehensive account here, we concentrate on some properties near the percolation 'multicritical' point at zero temperature and concentration p_c, where the diluted lattice is fractal.

The transition temperature comes to zero on this point (figure 5) because there the infinite cluster, of spins connected via the magnetic interaction, disappears. The form of the phase boundary near p_c can be obtained (Young & Stinchcombe 1976; Stinchcombe 1979a) from the scaling equations for the concentration p and an appropriate thermal parameter t. These typically take the form

$$p' = R(p), \quad t' = T(t, p). \tag{7}$$

An example of the derivation of the first equation was given in §3, and the second equation arises from scaling the distribution (which is initially binary) for the variable $K_{ij} = J_{ij}/K_B T$, where J_{ij} is the exchange constant (initially J or 0 depending on whether sites i, j are both present or not). In the first such treatment (Young & Stinchcombe 1976) it was found for the Ising system using the variable $t \equiv \exp[-2J/K_B T]$ that the eigenvalue λ ($= [\mathrm{d}p'/\mathrm{d}p]$ at p_c) of the linearized form of the first scaling equation (see §3) was also the eigenvalue resulting from the second equation ($[\mathrm{d}t'/\mathrm{d}t]$ at $t = 0$, $p = p_c$). That implies that $\exp[-2J/k_B T_c(p)]$ is linear in $p - p_c$ near p_c. For the Heisenberg systems, by using variable $t \equiv J/k_B T$, it can be seen (Stinchcombe 1979b) that the second scaling equation is the same as that for the scaling of the conductance of the percolation network. That implies that the dependence of T_c on $(p - p_c)$ near p_c involves as 'crossover exponent' the ratio t/ν, where t is the conductance exponent. These results for Ising and Heisenberg crossover at the multicritical point were subsequently established on a firm basis by Coniglio (1981), and agree with experiment (Cowley et al. 1977, 1980).

It turns out, however, that the fractal character of the percolation system near p_c leads there to another crossover in Heisenberg systems. This is because the excitation of spin waves affects such quantities as the magnetization, etc. and the fractal background makes the spin wave dynamics anomalous at p_c (§§5 and 6). This leads near p_c to a further, dynamically induced, crossover involving the spectral dimension (§§5 and 6), in magnetization (Korenblit & Shender 1978; Stinchcombe & Pimentel 1988), specific heat, susceptibility etc. (Pimentel & Stinchcombe 1989). The crossover is to anomalous temperature dependences (also field dependences) of such quantities, and should be experimentally accessible.

The theoretical discussion just referred to has implications for the existence or not of an ordered phase for Heisenberg spins on a typical fractal. If the (boson) fluctuations that can destroy the order have spectral dimension d_s, long-range order is impossible if $d_s \leqslant 1$. The form of statement just given applies also for loss of long-range positional order (Pal et al. 1987) and Bose–Einstein condensation (P. Pfeifer, personal communication) on typical fractals with $d_s \leqslant 1$. It is at present not clear, however, whether it can cover 'exceptional' cases (e.g. of fractals like the Sierpiński gasket with disconnecting nodes) of the sort discussed by Gefen et al. (1980).

5. DYNAMICAL PROPERTIES OF FRACTALS

The dynamics of dilute magnets near the percolation threshold will be discussed in §6. This is an example of dynamics on a statistically self-similar structure (random fractal) and in common with regular fractals the scale invariance leads to anomalous (or 'critical') dynamics. The discussion of dynamics of fractals that now follows, which is not restricted to the case of dilute magnets at p_c, is presented to provide a general background for the discussion in §6, and to put it into context.

Systems so far investigated in connection with anomalous dynamics on random fractals include resins and glasses (Rosenberg 1985), aerogels (E. Courtens, this symposium) and the dilute magnets for which details will be given later. In the resins, glasses and aerogels the excitations sought were the 'fractons' (Orbach 1985), which phonons become in a condition of self-similarity, i.e. inside the scaling window referred to in §1. In the magnetic systems, an analogous crossover (to be described subsequently) occurs for the magnons (spin waves) in Heisenberg magnets (Korenblit & Shender 1978; Stinchcombe 1985 b). However, the variety of dynamical processes available in magnets is very wide, encompassing the linear precessional dynamics of the spin waves but also relaxational dynamics: this can be purely diffusive but more usually involves also activation (which is a non-linear process as in Glauber or Kawasaki (conserved) spin flip dynamics for Ising or anisotropic Heisenberg models). Dramatically new anomalous dynamical behaviour can occur in this latter case (§6).

As well as the rich variety of dynamical processes, the dilute magnets have the advantage, shared with aerogels, that they can be prepared with a very large characteristic length ξ (i.e. with concentrations near p_c), (Cowley *et al.* 1977), so the system can behave, over a large scaling window, like a fractal.

To illustrate the effect of scale invariance on dynamics, and the sort of scaling techniques that can describe it, consider as an idealized example linear dynamics on the Sierpiński gasket (Rammal & Toulouse 1983; Harris & Stinchcombe 1983). We can discuss together ferromagnetic spin wave (magnons), scalar phonons, and diffusion by defining a variable Ω equal to ω, ω^2 or $i\omega$ in the three cases respectively, where ω is the frequency. Then the equations of motion are the same for the three cases. Moreover, under an exact 'decimation' transformation removing the sites shown as open circles in figure 1, i.e. working backwards along the hierarchy shown in the figure, the equations of motion are isomorphic with only the parameter change

$$\Omega \to \Omega' = \lambda\Omega - \Omega^2, \tag{8}$$

where $\lambda = d + 3$ for a hypertetrahedral Sierpiński gasket in d euclidean dimensions (so $\lambda = 5$ for the triangular case shown in the figure). At the same time the lattice spacing a scales according to $a \to a' = ba$, where $b = 2$. These statements imply that, for small Ω (where the quadratic terms in (8) can be discarded)

$$\Omega \propto a^z, \tag{9}$$

where $\lambda = b^z$, so $z = \lg_2 (d+3)$. Thus (for the interesting cases, $d \geqslant 2$) the length scaling dimension z of the dynamic parameter Ω is not 2, as it would have been in a non-fractal system in which Ω is proportional to $(ka)^2$ where k is the wave vector (as results from introducing the normal quadratic (spin waves) or linear (phonons) dispersion relation between ω and k into the definition of Ω). Instead the dynamics is anomalous or critical, having 'dynamic exponent' $z > 2$.

For this idealized example, equation (8) allows (Domany *et al.* 1983) the extraction of the density of states, which is a Cantor set because of the repeller arising, for $\lambda > 4$. At low Ω the integrated density of states $N(\Omega)$ behaves like

$$N(\Omega) \propto \Omega^{d_s}, \quad d_s = d_t/z, \tag{10}$$

where d_s is the 'spectral dimension' (Alexander & Orbach 1982, Rammal & Toulouse 1982). It satisfies $d_s = d_t/z$, because the fractal dimension d_t is the length scaling dimension of N. This is a general result. The full dynamic structure factor $R(\Omega k)$, which is the response of the gasket to a dynamic probe of frequency Ω and wave vector k, has also been obtained, by combining (8) with a scaling equation for R (Maggs & Stinchcombe 1986). For small, Ω, k the result is a function only of Ω/k^z, which means it has the 'dynamic scaling' form proposed in the context of critical dynamics by Hohenberg & Halperin (1977); except that there is in addition a (periodic) dependence on $\lg_2 k$ arising from the discrete scale invariance of this regular fractal (see figure 6). Such features (which also arise in the density of states) make regular fractals of this type less good models of the percolation infinite cluster when discussing dynamical properties than in simpler static structures.

Turning next to real systems, these normally have a finite, but possibly large, characteristic length ξ. The anomalous dynamics ($\Omega \propto k^z$) with exponent $z > 2$

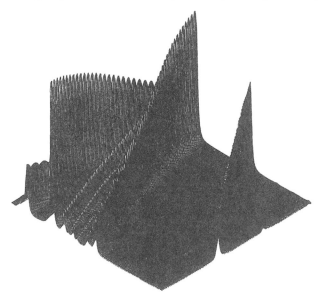

FIGURE 6. Dynamic structure factor for linear dynamics on the Sierpiński gasket. The ordinate is the response of the system to a probe of frequency Ω and wave vector k. The abscissae are Ω/k^z (as suggested by dynamic scaling) and k.

only applies when ξ is greater than the excitation or probe length $1/k$, that is, for $k\xi \gg 1$. For $k\xi \ll 1$ (which is equivalent to ξ much smaller than the wavelength), the excitation sees the system as homogeneous and the dynamics becomes normal, i.e. Ω is quadratic in k. According to the dynamic scaling hypothesis proposed in the context of critical dynamics at phase transitions by Hohenberg & Halperin 1977),

$$\Omega = k^z f(k\xi) \rightarrow k^z, \; k\xi \gg 1 \left.\right\} \atop \rightarrow Dk^2, \; k\xi \ll 1 \right\} \tag{11}$$

(all other lengths in the system being regarded as much smaller than ξ or $1/k$). This includes as shown the two limiting forms, anomalous and normal (i.e. critical and hydrodynamic), discussed above.

It can be seen from (11) that the coefficient D must be proportional to ξ^{2-z}, and as $z > 2$ and ξ diverges at the transition D must tend to zero at the transition. This is the 'mode-softening' phenomenon familiar in critical dynamics. However, as the transition is approached and ξ becomes larger and larger, eventually $k\xi$ becomes of order unity and the Dk^2 behaviour crosses over to the k^z anomalous dynamics. For phonon dynamics in systems where ξ large is associated with a real fractal (e.g. aerogels, percolation, etc.), the excitation associated with the k^z anomalous dynamics has been called the 'fracton' (Orbach 1985). Obviously analogous effects occur in the magnon dynamics of diluted magnets as $p \rightarrow p_c$ (§6).

The dynamic exponent z can be related to static exponents by the following sort of argument, which we now give for the specific case of spin waves. As seen above, $2-z$ is the length-scaling exponent of the coefficient D, which is the same as the diffusion constant. This is related by an Einstein relation $D = \sigma/n$ to the conductance σ and density n. Therefore, $z-2$ is the difference of the length scaling exponents of n and σ. For the infinite cluster in diluted systems near the percolation threshold the density is $n = P(p)$ whose length scaling exponent was seen in §2 to be $-\beta/\nu$. Similarly that for the conductivity is t/ν, so

$$z = 2 + (t-\beta)/\nu \equiv z_F. \tag{12}$$

This equation expresses the dynamic exponent (for ferromagnetic spin waves on the infinite cluster in dilute magnets near p_c) in terms of static exponents (Korenblit & Shender 1978; Harris & Stinchcombe 1983). For the antiferromagnetic case, where spin waves have a linear dispersion in the normal régime, a similar argument can be applied by using a result of Harris & Kirkpatrick (1977). This results in (Christou & Stinchcombe 1986a)

$$z_A = 1 + \tfrac{1}{2}(t-\tau)/\nu, \tag{13}$$

where τ is the exponent for the static transverse susceptibility. By using approximate (but accurate) relations for the percolative exponents, (13) becomes $z_A \sim \tfrac{2}{3}z_F \sim d_f$, unlike the result $(z_A = \tfrac{1}{2}z_F)$ found on simple regular fractals (Stinchcombe & Maggs 1986). Note that z_A is not the same as the scalar phonon dynamic exponent despite the fact that normal phonon dynamics also has a linear dispersion, because the stiffnesses of the two problems scale differently.

Knowledge of z and the fractal dimension d_f gives, by virtue of (10), the spectral dimension. Alternatively, this can be obtained by direct numerical techniques (Lewis & Stinchcombe 1984). The β-exponent for localization can be expressed in terms of d_s by using arguments like those above, with the conclusion that fractons are localized (Rammal & Toulouse 1983). As referred to at the end of §4, d_s is the dimension characterizing the new thermal crossover effects (in spin wave magnetism, etc.) induced by the anomalous dynamics.

Hereafter we turn specifically to the case of anomalous spin dynamics on the percolation fractal in dilute magnets. This will involve both linear dynamics, as described, and nonlinear dynamics.

6. CRITICAL DYNAMICS IN DILUTED MAGNETS NEAR THE PERCOLATION THRESHOLD

In this discussion of anomalous dynamics in diluted spin systems near p_c we begin with the (linear) dynamics of spin waves on the percolation background, and in a second subsection consider nonlinear dynamics.

(a) Spin wave dynamics near p_c

The previous section gives, in a general context, the main theoretical results for *linear dynamics* (normal to critical crossover, etc.) and specific results for exponents. More detailed results, e.g. for crossover functions (e.g. f in (ii)), density of states functions, dynamic response function, spin wave widths etc. can, in principle at least, be obtained by scaling variables like Ω and p for the random system (cf. (7) for statics). Some examples are given in Stinchcombe (1983b), Harris & Stinchcombe (1983), Maggs & Stinchcombe (1986), and these studies are continuing. In addition, exact results can be obtained for diluted one-dimensional systems, e.g. for the full dynamic structure factor (Stinchcombe & Harris 1983; Maggs & Stinchcombe 1984, Maggs *et al.* 1986). Alternatively, standard techniques for the normal régime can be combined with crossover arguments, e.g. to discuss spin wave damping (Christou & Stinchcombe 1986a, b). In this problem, a Rayleigh scattering analysis in the normal régime but allowing (through crossover) for the large characteristic scale of the background for large ξ, yields the following dynamic scaling forms and asymptotic results for the ferro- and antiferromagnetic spin wave widths Γ_F, Γ_A:

$$
\left.
\begin{aligned}
\Gamma_F = k^{z_F} g_F(k\xi) &\to k^{z_F} & k\xi \gg 1 \\
&\to \xi^{d+2-z_F} k^{d+2} & k\xi \ll 1 \\
\Gamma_A = k^{z_A} g_A(k\xi) &\to k^{z_A} & k\xi \gg 1 \\
&\to \xi^{d+1-z_A} k^{d+1} & k\xi \ll 1
\end{aligned}
\right\}
\tag{14}
$$

Neutron-scattering results are now beginning to provide clear evidence of anomalous dynamical effects produced by critical dilution in dilute magnets. An example is the work of Uemura & Birgeneau (1987) on spin wave dynamics in the Heisenberg antiferromagnet $Mn_p Zn_{1-p} F_2$. Another experiment, on nonlinear spin dynamics in a diluted Ising system, will be referred to in subsection (b).

Uemura & Birgeneau (1987) measured spin wave energies ω and widths Γ at two different concentrations. At the first, well away from p_c, Γ was much less than ω for all wave vectors k (underdamped situation). At the second concentration, closer to p_c, the width was seen to exceed the energy for wave vectors k beyond some value k^* (figure 7). This is the behaviour expected as one moves, by increasing $k\xi$, from the normal to the anomalous régime. For this interpretation to be correct, k^* should be about $1/\xi$. Direct measurement of $1/\xi$ confirms that it is so, providing strong evidence that the crossover scaling behaviour (to the spin wave analogue of the fracton) has been seen in this dilute Heisenberg (antiferro) magnet. The crossover should not occur at the first concentration because there the correlation length is not long enough to give a scaling régime.

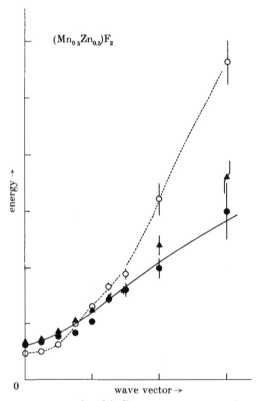

FIGURE 7 Spin wave energy ω and width Γ against wave vector k as measured by inelastic neutron scattering in a dilute Heisenberg magnet near the percolation threshold (Uemura & Birgeneau 1987). The crossing of the two curves is interpreted as the crossover from normal to anomalous ('fracton') dynamics.

The experimental results for ω, Γ against k above the crossover region for the concentration closest to p_c fit reasonably well to straight lines on logarithmic plots. However, the slopes are about 1.05 ± 0.3 and 1.55 ± 0.2 respectively, whereas the expressions (11), (14) predict that both slopes should be $z = z_A \sim 2.5$, using (13) for the three-dimensional isotropic antiferromagnet. An anisotropy gap is apparent in the experimental data, so it would be more appropriate to fit to a theory allowing for anisotropy effects.

(b) *Ising dynamics at the percolation threshold at low temperature*

Ising systems have nonlinear relaxational dynamics in which spin flips are induced by contact with a heat bath (typically the phonons). Unlike the linear spin wave dynamics the temperature has to be non-zero for any relaxation, so thermal factors enter the dynamics. At the percolation threshold one might therefore expect the self similarity of the background to result in the following relaxational analogue of the anomalous dynamics just discussed, in which ω is replaced by the inverse of the relaxation time τ, and k by $1/\xi_T$ where ξ_T is the thermal correlation length:

$$\tau \propto \xi_T^z. \tag{15}$$

However, theoretical work (Harris 1983; Henley 1985; Stinchcombe 1985b; Harris & Stinchcombe 1986) leads to a much more singular relation

$$\tau \propto \xi_T^{A\ln\xi_T + B}, \tag{16}$$

which if cast into the form (15) involves a diverging effective exponent. We first indicate the reason for this (following Harris & Stinchcombe 1986) and then describe experimental work and simulation.

The relaxation is by domain wall diffusion over the percolation fractal. This in common with the backbone shown in figure 2, contains linear sections ('chains') over which diffusion is rapid. However, the process of moving the domain wall over a node at which a chain branches into two chains costs energy so the rate is slowed by an activation factor, $R = \exp(-2J/k_BT)$). Because the geometry of nodes and links is self-similar (i.e. there is a hierarchy of branchings on branchings) length scaling of the relaxation time τ involves a multiplicative activation factor

$$\tau' \sim \frac{1}{R}\tau \sim (\xi_T/a)^{1/\nu}\tau. \tag{17}$$

In this expression R has been written in terms of ξ_T, and the percolation exponent ν (using the result, §4, that the Ising thermal/percolation crossover exponent is unity). ξ_T diverges at the (zero temperature) fixed point, and for that reason all other scaling factors are unimportant and have been omitted. So (17) does not have a finite eigenvalue (unlike, for example, λ in (8)) and the usual procedures do not apply. However, the scaling equations for $\ln\tau$ and $\ln(\xi_T/a)$ can be easily combined to form a scaling invariant, and this leads simply to $\ln\tau = A(\ln\xi_T)^2 + B\ln\xi_T$, which is equivalent to (16).

The discussion can be generalized in various ways. For instance, (16) applies for the asymptotic relaxational dynamics in anisotropic Heisenberg magnets (Christou & Stinchcombe 1989) or in Potts models (Jain et al. 1986). Further, Rammal (1987) has proposed a generalization based on the invariance of energy-barrier distributions under simple composition laws which may have much wider applications.

An experimental investigation of Ising dynamics at low temperatures and $p \sim p_c$ has been carried out by Aeppli et al. (1984). The slow relaxation was measured by inelastic neutron scattering on the two-dimensional Ising anti-ferromagnet $Rb_2Co_pMg_{1-p}F_4$, with p so close to p_c that ξ_p was hundreds of lattice spacings. The resulting logarithmic plot of reduced inverse relaxation time against reduced inverse thermal correlation length (κa) is shown in figure 8. The straight line is a fit to (15), and leads to a large exponent z. A possible explanation of the fact that the curvature implied by (16) is not seen is that relatively short thermal correlation lengths are involved in the experiment ($(1/\kappa a) \lesssim 3$). According to (16), the line in the logarithmic plot joining the experimental points should curve down (parabolically) to greater and greater slope if the experiment is extended to small enough values of κa, as has been planned (Aeppli 1986).

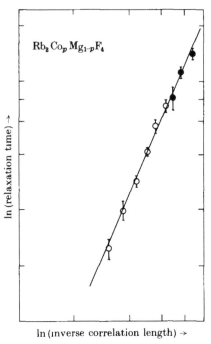

FIGURE 8. Anomalously slow relaxation in an Ising layer magnet at the critical concentration p_c (Aeppli *et al* 1984). Experimental results are shown in a logarithmic plot of reduced inverse relaxation time against reduced inverse correlation length.

In the meantime the result (16) has been subjected to tests by Monte Carlo simulation. Although the excessively slow dynamics makes such investigations very difficult, the published results (Jain 1986a, b; Chowdhury & Stauffer 1986) and one unpublished one (C. K. Harris & D. Nicolaides, personal communication) give clear evidence for the behaviour (16) at the longest thermal correlation lengths available. In addition, indirect support is obtained from energy barrier calculations (Rammal & Benoit 1985a, b).

The critical dynamics of dilute magnets near the percolation threshold is worthy of much further study, experimentally and theoretically, and it is one of the richest areas for real investigations of dynamics on fractals.

REFERENCES

Aeppli, G. 1986 *Physica* B, C **136**, 301.
Aeppli, G., Guggenheim, H. J. & Uemura, Y. J. 1984 *Phys. Rev. Lett.* **52**, 942.
Alexander, S. & Orbach, R. 1982 *J. Phys. Lett.* **43**, L625.
Barber, M. N. 1975 *J. Phys.* C **8**, L203.
Broadbent, S. R. & Hammersley, J. M. 1957 *Proc. Camb. phil Soc* **53**, 629.
Chowdhury, D. & Stauffer, D. 1986 *J. Phys.* A 19, L19
Christou, A. & Stinchcombe, R. B. 1986 *a J. Phys.* C **19**. 5917.
Christou, A. & Stinchcombe, R. B. 1986 *b J. Phys.* C 19, 5895.
Christou, A. & Stinchcombe, R. B. 1989 *Phys. Rev.* B (In the press.)
Coniglio, A. 1981 *Phys. Rev. Lett.* **46**, 250.
Cowley, R A., Shirane, G., Birgeneau, R. J. & Svensson, E. C. 1977 *Phys. Rev. Lett.* **39**, 894.
Cowley, R. A., Birgeneau, R. J., Shirane, G., Guggenheim, H. J. & Ikeda, H. 1980 *Phys Rev.* B **21**, 4038.
Domany, E., Alexander, S., Bensimon, D. & Kadanoff, L. P. 1983 *Phys. Rev.* B **28**, 3110.
Essam, J. W. 1980 *Rep. Prog. Phys.* **43**, 83.
Gefen, Y., Mandelbrot, B. & Aharony, A. 1980 *Phys. Rev. Lett.* **45**, 855.
Harris, A. B. & Kirkpatrick, S. 1977 *Phys. Rev.* B **16**, 542.
Harris, A. B., Lubensky, T. C., Holcomb, W. K. & Dasgupta, C. 1975 *Phys. Rev. Lett.* **35**, 327.
Harris, C. K. 1983 *D. Phil. Thesis*, Oxford University.
Harris, C. K. & Stinchcombe, R. B. 1983 *Phys. Rev. Lett* **50**, 1399.
Harris, C. K. & Stinchcombe, R. B. 1986 *Phys. Rev. Lett.* **56**, 896.
Henley, C., 1985 *Phys. Rev. Lett.* **54**, 2030.
Hofstadter, D. R. 1976 *Phys. Rev.* B **14**, 2239.
Hohenberg, P. C. & Halperin, B. I. 1977 *Rev. mod. Phys.* **49**, 435.
Jain, S. 1986 *a J. Phys.* A **19**, L57.
Jain, S. 1986 *b J. Phys.* A **19**, L667.
Jain, S., Lage, E. J. S. & Stinchcombe, R. B. 1986 *J. Phys.* C 19, L805.
Kadanoff, L. P. 1966 *Physics* **2**, 263.
Kadanoff, L. P. & Houghton, A. 1975 *Phys. Rev.* B 11, 377
Korenblit, I. Y. & Shender, E. F. 1978 *Usp. fiz. Khim.* **126**, 233. (*Soviet Phys. Usp.* **21**, 832 (1978).)
Lewis, S. J. & Stinchcombe, R. B. 1984 *Phys. Rev. Lett.* **52**, 1021.
Ma, S. K. 1976 *Modern theory of critical phenomena.* New York: Benjamin.
Maggs, A. C. & Stinchcombe, R. B. 1984 *J. Phys.* A **17**, 1555.
Maggs, A. C. & Stinchcombe, R. B. 1986 *J. Phys.* A **19**, 2637.
Maggs, A. C., Gonçalves, L. L. & Stinchcombe, R B. 1986 *J. Phys.* A **19**, 1927
Mandelbrot, B. 1977 *Fractals.* San Francisco · Freeman.
Niemeyer, T. L. & Van Leeuwen, J. M. J. 1974 *Physica* **71**, 17.
Orbach, R. 1985 In *Scaling phenomena in disordered systems* (ed. R. Pynn & A. Skjeltorp), pp. 335–359. New York. Plenum.
Pal, B., Manna, S. S. & Chakrabarti, B. K. 1987 *Solid St. Commun.* **64**, 1309.
Pimentel, I. R. & Stinchcombe, R. B. 1989 (In the press.)
Rammal, R. 1987 In *Time dependent effects in disordered materials* (ed. T. Riste & A. Skjeltorp). New York: Plenum.
Rammal, R. & Benoit, A. 1985 *a Phys. Rev. Lett.* **55**, 649.
Rammal, R. & Benoit, A. 1985 *b J. Phys. Lett.* **46**, 667.
Rammal, R. & Toulouse, G. 1982 *Phys. Rev. Lett.* **49**, 1194.
Rammal, R. & Toulouse, G. 1983 *J. Phys. Lett.* **44**, L13.
Reynolds, P. J., Klein, W. & Stanley, H. E 1977 *J. Phys.* C **10**, L167.
Rosenberg, H. M. 1985 *Phys. Rev. Lett.* **54**, 704.
Stanley, H. E. 1985 In *Scaling phenomena in disordered systems* (ed. R. Pynn & A. Skjeltorp), pp. 49–69. New York. Plenum.
Stinchcombe, R. B. 1979 *a J. Phys.* C **12**, 4533.
Stinchcombe, R. B. 1979 *b J. Phys.* C **12**, 2625.

Stinchcombe, R. B. 1983*a* In *Phase transitions and critical phenomena* (ed. C. Domb & J. L. Lebowitz), vol. 7, p. 151 New York. Academic Press.
Stinchcombe, R. B. 1983*b* *Phys. Rev. Lett.* **50**, 200.
Stinchcombe, R. B. 1985*a* In *Scaling phenomena in disordered systems* (ed. R. Pynn & A. Skjeltorp), pp 13–30 New York Plenum.
Stinchcombe, R. B 1985*b* In *Scaling phenomena in disordered systems* (ed. R. Pynn & A. Skjeltorp), pp. 465–482. New York · Plenum.
Stinchcombe, R B. & Harris, C. K. 1983 *J. Phys.* A **16**, 4083
Stinchcombe, R B. & Maggs, A. C. 1986 *J. Phys.* A **19**, 1949.
Stinchcombe, R. B. & Pimentel, I. R. 1988 *J. Phys.* A **21**, L807.
Uemura, Y. J & Birgeneau, R. J. 1987 *Phys. Rev.* B **36**, 7024.
Wilson, K. G. 1972 *Phys. Rev. Lett.* **28**, 584.
Wilson, K G & Fisher, M. E. 1972 *Phys Rev Lett.* **28**, 240
Young, A. P. & Stinchcombe, R. B. 1975 *J. Phys.* C **8**, L535.
Young, A. P. & Stinchcombe, R. B. 1976 *J. Phys* C **9**, 4419

Discussion

E. Courtens (*IBM Research Division, Rüschlikon, Switzerland*). In the published work of Uemura & Birgeneau, there is no determination of a spectral dimension d_s. Is Dr Stinchcombe aware of an experimental determination of d_s in $Rb_2Co_pMg_{1-p}F_4$ or another crystal of that family?

R. B. Stinchcombe. Critical dynamics has been measured at the percolation threshold in the system $Rb_2Co_pMg_{1-p}F_4$ that Dr Courtens asks about (Aeppli *et al.* 1984). This does provide an accurate value (in a particular temperature range) for a dynamic exponent z (rather than d_s). However, this system is a layer *Ising* magnet, with nonlinear relaxational dynamics; so simple fracton viewpoints do not apply and instead the anomalous dynamics should take the highly singular form described briefly at the end of the lecture. In particular the measured z should be temperature dependent, according to the theory. The published work of Uemura & Birgeneau (1987) is the only experiment so far known to me that measures dynamical properties in the fracton régime of the Heisenberg spin wave system. In principle it could provide a direct determination of the dynamic exponent z (and hence d_s, e.g. via (10) and (4)). However, the scaling window is not very wide, because the concentration is not very close to p_c, and anisotropy effects need to be allowed for.

Fractals and phase separation

By D. W. Schaefer, B. C. Bunker and J. P. Wilcoxon

Sandia National Laboratories, Albuquerque, New Mexico 87185, U.S.A.

Fractal patterns are typically observed in systems that develop far from equilibrium. Because phase separating systems are far from equilibrium following a quench, we searched for fractal correlations in two quenched systems: borosilicate glasses and micellar solutions. Because scattering techniques are used, we review the expected scattering profiles expected for spinodal decomposition, nucleation and growth, and kinetic aggregation.

In the borosilicate glasses, no non-classical structures are observed unless the boron-rich phase is leached out. Upon leaching, a peak appears in the scattering profile and, depending on leaching conditions, surface fractal porosity develops. We trace the observed structure to the leach process itself, which we believe produces phase separation by nucleation and growth. Short-range structure is explained in terms of reaction-limited dissolution.

In micelle solutions, we find fractal patterns following a quench along the critical isochore into the unstable régime. A quench of equivalent depth off the critical isochore into the metastable régime produces no unusual structures. Fractal scattering profiles are also observed in the single-phase régime near the critical point. These observations suggest growth of polymeric structures in the single-phase régime.

1. Introduction

Fractal patterns typically appear in systems that develop far from equilibrium. These patterns are often interpreted by kinetic models, the properties of which are usually known only through computer simulation. In their simplest form, kinetic models do not allow for structural rearrangement and therefore do not yield minimum energy configurations. By contrast, the more common thermodynamic models describe growth phenomena close to equilibrium. These models involve parameters for the surface and bulk energy, and predict compact non-fractal structures.

The traditional models for phase separation, nucleation and growth, and spinodal decomposition, are thermodynamic models. These models successfully describe the essential experimental observations in most systems. There is reason to believe, however, that unique, possibly fractal, patterns exist in the earliest stages of phase separation where kinetic growth processes are active. At least for deep quenches, initial growth must take place far from equilibrium, exactly the conditions where fractal correlations are expected.

For both simple fluids and binary mixtures, it is well known that spatial density or concentration fluctuations are accurately described by Ornstein–Zernike (O.Z.) theory in the single phase near the critical point. The Q.Z. correlation function, however, is power-law at short distances, corresponding to fractal correlations. It is reasonable, therefore, that fractal concepts may be useful to describe the transition from O.Z. fluctuations in the single-phase region to uniformly dense structures in the late stages of growth in the two-phase régime.

Computer simulation of spinodal processes reveal highly ramified clusters that are initially described by percolation theory. Grest & Srolovitz (1984), for example, studied the evolution of two-dimensional (2D) Ising clusters and found that the spanning percolation cluster first became more ramified (i.e. D, the fractal dimension, decreased) before the clusters ultimately became compact ($D \to 2$, the dimension of the embedding space). Desai & Denton (1986) reported qualitatively similar results. The microscopic observations of Guenoun et al. (1988) confirm ramified clusters in a binary fluid near the critical point.

The above considerations led us to seek systems in which we might observe the earliest stage of phase separation. Basically we seek power-law correlations in the structure factor measured by small-angle X-ray scattering (saxs) and small-angle neutron scattering (sans). Needless to say, we chose systems that optimize the possibility of observing fractal fluctuations. In addition to fundamental issues, we are also motivated by the possibility of creating new materials by exploiting the unusual patterns that emerge in non-equilibrium systems. To aid in the interpretation of the scattering data we first review, from the perspective of fractal geometry, the shape of scattering profiles from disordered systems.

2. FRACTALS

Figure 1a is a 2D, highly disordered system, phase separated into black and white phases. Analysis of this pattern on various length scales will help in understanding the shape of the scattering profiles observed in real systems.

On length scales comparable to the edge of figure 1a the system is uniformly dense and devoid of spatial correlations. Apart from near-critical systems, most systems are uniformly dense on length scales exceeding 1000 Å†.

On shorter length scales, disorder becomes apparent. Figure 1 (b–e), for example, shows figure 1a at four levels of magnification. On the scale of (b), one's eye is sensitive of the correlation range ξ, the largest length on which the system is non-uniform.

The magnification is enhanced further in figure 1c. Dominant fluctuations are now short compared to ξ, but long compared to any lattice parameter or bond length. If concentration fluctuations decay in power-law fashion, the geometry of the system can be described as mass-fractal on this intermediate length scale. Diluted gels, and colloidal aggregates fulfill this criterion and are characterized by a mass-fractal dimension D. In addition, one could imagine a two-phase system where ramified mass-fractal domains existed. A diluted percolation system is an example.

On further magnification the surface features of the domains are evident as in figure 1d. Surface roughness can also be described by fractal geometry. The surface fractal dimension D_s quantifies roughness. If $D_s = 2$, the surface is smooth

† $1 \text{ Å} = 10^{-10} \text{ m} = 10^{-1} \text{ nm}$.

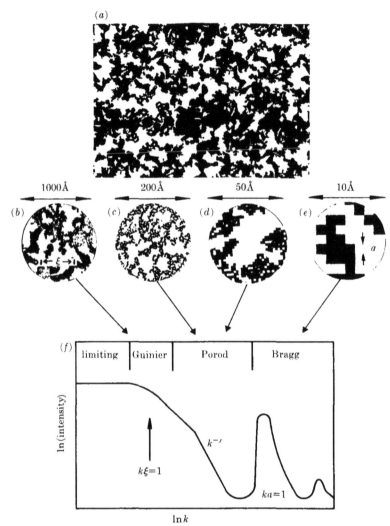

FIGURE 1. Schematic representation of a phase separated material. (a) Real space two-dimensional pattern. (b) The pattern of (a) displayed at a level of magnification where the overall cluster-size distribution is the dominant geometric feature. (c) On further magnification, the mass distribution within clusters appears. At this level, the mass fractal dimension D is useful to characterize the structure of the domains. (d) With further magnification. the structure of the domain interface emerges. Surface roughness can be characterized through the surface fractal dimension D_s. (e) At the highest level of magnification, the lattice structure emerges. (f) The scattering profile contains all the information described above. In the limiting régime. the curve is flat, indicative of uniformity on large length scales. The position of the break in the Guinier régime is determined by the correlation range ξ. The break occurs at ξ^{-1}. Fractal geometry is useful when power-law scattering profiles are found in the Porod régime. Both D and D_s can be extracted from the measured porod slopes. In the Bragg régime, atomic level correlations give rise to peaks.

in a 3D embedding space, and the surface area of a domain of size r scales as r^2, as expected for conventional euclidean objects. For fractally rough domains, however, the surface area can scale with any exponent lying in the interval $2 < D_s < 3$.

The mass and surface fractal descriptions, of course, only apply to systems that are self-similar or self-affine (in the sense described by Mandelbrot (1982)). From the view of figure 1, self-similarity means that, for a limited range of magnification power, the system looks the same.

Except for the singular case $D = D_s = d$, objects cannot be both mass-fractal and surface-fractal on the same length scale. For a mass-fractal, the surface area scales with mass whereas for surface fractals it scales as mass to the power $\frac{1}{3}D_s$. Some systems, like colloidal aggregates (Hurd et al. 1987), can display both characteristics on different length scales.

Figure 1e reveals that the system lies on a square lattice. In a real system, of course, correlations at this level correspond to atomic level structure. Although atomic level forces underlie phase separation, they are of secondary interest here as the long-range structure is probably independent of the details of interatomic or intermolecular potentials.

The system in figure 1 could be characterized in real space by quantifying the analysis presented above. All of the information discussed (ξ, D, D_s, lattice structure) could be extracted from the pair correlation function. Alternatively, D and D_s could be taken from logarithmic plots of domain mass and surface area against length. These real space procedures are quite successful in two dimensions and were used by Guenoun et al. (1987) in their studies of phase separation. In 3D systems, however, overlap and shadowing considerably complicate real space analysis. For this reason, scattering methods are preferred.

Scattering methods measure the Fourier transform of the pair correlation function and provide the same information discussed above. By measuring the scattered intensity as a function of wave vector k, one effectively obtains information on length scales corresponding to k^{-1}, as indicated schematically in figure 1f. As $k \to 0$ the scattered intensity I becomes independent of k, consistent with the spatial uniformity of (a) on the scale of the edge length. The break in the Guinier region occurs at $k = \xi^{-1}$, and the slopes in the Porod régime depend on D and D_s (Schaefer 1988). Finally, peaks are observed in the Bragg régime, the position of which depends on short-range atomic correlations.

We are primarily concerned with the Porod régime where the power-law exponent P for the decay of $\ln I$ against $\ln k$ is

$$P = -2D + D_s. \tag{1}$$

For mass fractals, surface area scales with mass so $D = D_s$ and $P = -D$. That is, the mass fractal dimension is the slope of $\ln I$ against $\ln k$. Surface fractals, by contrast, are uniformly dense so $D = d = 3$ and $P = -6 + D_s$. Often a 'crossover' separates the two régimes as indicated by the break in the Porod régime of figure 1f. The crossover length is k^{-1} at the break.

Using the Porod slopes, we identify mass and surface fractals as materials with P greater than and less than -3 respectively, assuming of course that power-law analysis is justified. In situations where interfacial layers are present, scattering profiles with $P < -4$ are found, corresponding to $D_s < 2$.

Based on the above discussion, we can now predict the development of the scattering profiles for various models of phase separation. For nucleation and growth, small, uniformly dense $(D = 3)$ smooth surfaced $(D_s = 2)$ domains form and grow in size. Figure 2 shows the expected development of the scattering profile assuming a conserved order parameter (large domains growing at the expense of small). The mass of the domains (proportional to the intercept at $k = 0$) increases but the surface area (proportional to I in the power-law régime) decreases.

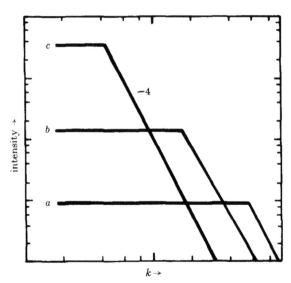

FIGURE 2 Development of the scattering profile for nucleation and growth. At the earliest times (line *a*) small smooth surfaced particles appear. Later (lines *b* and *c*), large particles grow at the expense of small ones. The break in the curve (Guinier régime) moves to smaller *k* as the clusters grow. Simultaneously, the amplitude in the Porod régime decreases because the surface area of the system is decreasing.

In spinodal decomposition (figure 3) a dominant unstable Fourier component appears in the spatial fluctuation spectrum. This component produces a peak in the scattering profile that moves to smaller k, corresponding to longer length scales, as phase separation proceeds. The high-k régime is expected to start with a slope of -2 (the O.Z. value) and approach -4 as distinct domains form.

Kinetic growth should give a fundamentally different profile as indicated in figure 4. Now, the minority phase grows by a kinetic process such as diffusion-limited aggregation. This process gives rise to mass fractals with large clusters forming from smaller ones. Eventually, of course, kinetic growth should crossover to one of the thermodynamic régimes described above. Presumably, short-scale ramification would anneal out first giving rise to slopes approaching -4 at large k. The scattering profiles would steepen to -4 at high k and eventually the steeper profile should invade all of k space down to $k^{-1} = \xi$.

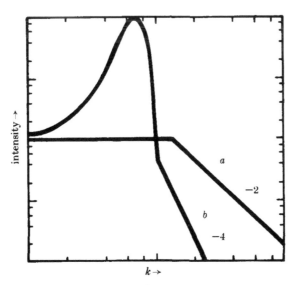

FIGURE 3. Scattering profile for spinodal decomposition. Line a is the O.Z. profile that characterizes fluctuations in the single-phase régime. As phase separation proceeds a peak develops and distinct domains form, eventually with sharp interfaces giving rise to limiting Porod slopes of -4.

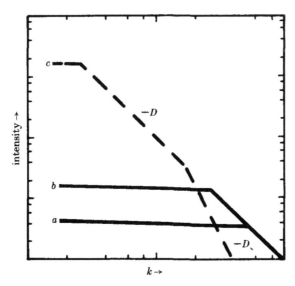

FIGURE 4. Scattering profile for kinetic growth. Aggregation gives rise to ramified domains (Porod slope $= -D$). Eventually (c) ripening takes place of short length scales leading to a transition in the Porod régime to steeper slopes characteristic of either smooth or fractal surfaces.

Although the above analysis is obviously oversimplified, hopefully it contains sufficient insight to make progress in the understanding of the development of structure in phase-separated systems. To illustrate this point, we consider two systems: borosilicate glasses and micellar solutions. For different reasons each system is a likely candidate for observing fractal patterns in the early stage of phase separation.

3. Borosilicate glass

Because kinetic growth processes occur in systems that are far from equilibrium, they are more likely to be found in glasses than in crystalline materials. In addition, the viscosity of network-formers is high even in the liquid state, so rapid quenches are likely to trap non-equilibrium structures that develop in the early stages of phase separation. Because of these effects, we studied the structure of a series of sodium borosilicate glasses, which spanned the phase diagram from strictly single phase (30–10–60, corresponding to the mole percentages of Na_2O, B_2O_3 and SiO_2) to strongly phase separated (5–35–60). Here we report on two compositions where phase separation is a factor in the structure.

The borosilicates are the basis for the preparation of Vycor porous glass. Because the boron-rich phase is soluble in acid, it can be extracted by leaching. In addition to the homemade samples, we also studied commercial Vycor 7930 obtained from Corning Glass Inc. (Corning N.Y., U.S.A.). The homemade samples were studied by SAXS and SANS in both the bulk and leached state whereas the commercial sample was studied only in the leached condition.

Figure 5 shows a comparison of the SANS and SAXS profiles for Vycor 7930. Apart from details, the curves look strikingly similar to the schematic curve for spinodal decomposition in figure 2. Indeed, this similarity prompted Schaefer *et al.* (1987)

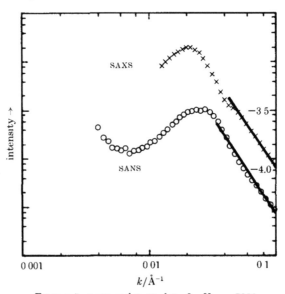

FIGURE 5. SAXS and SANS data for Vycor 7930.

to identify spinodal decomposition as the essential process that controls porosity in these materials. The differences in the curves are primarily due to the fact that the sample size is 10000 times larger for the SANS experiment so the structure present in the SAXS data is washed out by sample inhomogeneity. SAXS data taken in line geometry (the SAXS data in figure 5 were measured in pinhole collimation) is similar to the SANS data showing a mean slope of -4 (Schaefer *et al*. 1987). The X-ray data of Schmidt *et al*. (1987) look similar to our neutron data, whereas the neutron data of Sinha *et al*. (1987) look similar to our X-ray data. These minor sample-dependent curiosities do not alter any conclusions drawn below.

Hohr *et al*. (1988) noted that the peak in figure 5 could also be explained by nucleation and growth if one imagines densely packed monosized particles as might arise from a deep quench into the metastable régime.

The controversy concerning the origin of the peak in figure 5 is a reinactment of the same dispute that arose two decades ago based on electron microscopy data on leached glasses. Cahn (1965) actually developed his random spinodal theory to explain microscopic observations on Vycor. Haller & Macedo (1968), on the other hand, demonstrated that coalesced particles could give an interconnected structure similar to Cahn's random spinodal patterns. These authors also demonstrated the existence of uncoalesced particles in quenched samples that were subsequently nucleated near the liquidus.

The above models contrast with that proposed by Bunker *et al*. (1986). These authors believe that the porous structure is caused by the leaching process itself by nucleation of either particles or liquid-filled droplets (Bunker *et al*. 1984), by a dissolution repolymerization process. The nucleated domains then coalesce much like Haller's model for phase separation in the melt.

In an attempt to clarify the origin of porosity in leached borosilicates, we studied these material in both the bulk and leached state. Figure 6 compares the SAXS and SANS data for 10–30–60 in bulk unleached (lower curve) and acid-leached state (upper two curves). The data on the leached sample are similar to figure 5, suggesting similar preparation. Interestingly, there is no evidence for phase separation in the bulk samples: the material is clear and no peak is seen in the SAXS profile. Only weak scattering with a slope close to -4 is observed, indicative of scattering from nearly smooth surfaced regions (probably trapped bubbles). As sufficient contrast should exist to see phase separation by SAXS, we conclude that the 10–30–60 precursor glasses are not phase separated. This conclusion supports the idea that phase separation occurs during leaching.

We also studied a glass of composition 5–35–60. This sample is cloudy and obviously phase separated. In this case we observe quite intense small-angle scattering (figure 7) with a Porod slope of -4. Because the scattered intensity is a factor of 100 larger than that observed in the 10–30–60, we conclude that it arises from phase separation, not bubbles. Because the correlation range in these materials is about 1000 Å (compare the electron micrograph in figure 8 with that from Vycor 7930 in figure 9), we do not expect a peak in figure 7. The leached sample shows increased scattered intensity, but no change in profile (upper curve, figure 7). We conclude that the pore structure is a skeleton of the phase-separated parent material. This conclusion, however, cannot be drawn for the 10–30–60

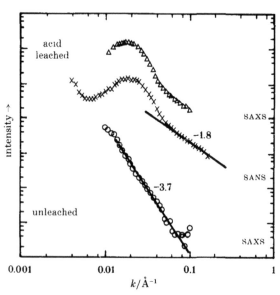

FIGURE 6. SAXS and SANS data for 10–30–60 sodium borosilicate glass in the leached and unleached state Samples were leached at pH = 1 for three days at 70 °C. Curves are not normalized with respect to each other The scattered intensity for the unleached glass is down by several orders of magnitude compared to the leached material

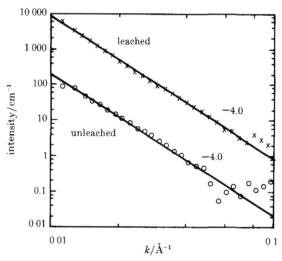

FIGURE 7. SAXS data for 5–35–60 sodium borosilicate glass Data are on an absolute scale so the scattering per unit volume increases by two orders of magnitude on leaching. Sample was leached at pH = 1 at 70 °C for four days.

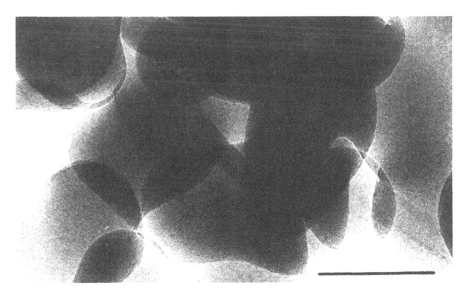

FIGURE 8. Transmission electron micrograph of leached 5–30–60 (from Bunker *et al.* 1986). Sample leached at pH = 8.5. Scale bar represents 0.1 μm.

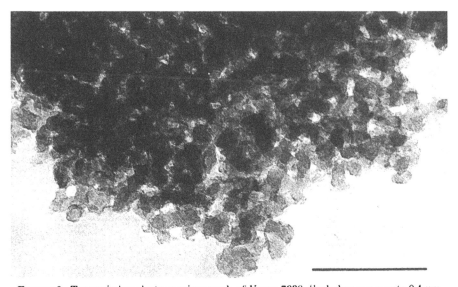

FIGURE 9. Transmission electron micrograph of Vycor 7930. Scale bar represents 0.1 μm.

sample where there is a qualitative difference between the bulk and leached glasses and where scattering in bulk is very weak.

At least for our 10–30–60 material, our results favour the idea that the essential structure of the leached material is determined by the leaching process, not phase separation in bulk. Indeed, Schaefer *et al.* (1988) already proposed that the high-k

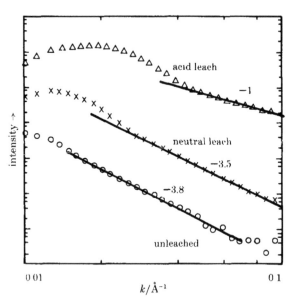

FIGURE 10 ᔕᴀxs data for 10–30–60 leached with acid and distilled water. Acid-leached data are the same as figure 6. The neutral leach was at 70 °C for two days.

régime of the ᔕᴀxs data for Vycor could be explained by a swollen leach layer dissolved by reaction-limited dissolution, a process believed to produce fractally rough surfaces. It must be realized, however, that the preparation of Vycor 7930 may involve ageing of quenched samples, a process that could lead to phase separation. Although our results do not preclude the possibility that the pore structure is the skeleton of phase separation in bulk, they suggest that phase separation in bulk is not necessary to produce scattering curves with 'spinodal peaks'.

Several explanations exist to account for the shape of the scattering curve at intermediate length scales corresponding to the region of k space just to the right of the peak. Wiltzius *et al.* (1987) suggest that higher harmonics of the basic spinodal frequency leads to a second peak, consistent with Berk's (1987) theory. Transmission electron microscopy (TEM) evidence on alkali silicate glasses (Bunker *et al.* 1984) however, indicates that nucleation and growth is the dominant process induced by leaching, not spinodal decomposition. Their data clearly show nucleation of particulate structures. Although similar evidence is lacking the borosilicates, the final pattern for all both glasses is similar.

Schaefer *et al.* (1988) proposed a swollen leach layer to explain the unusually steep slope just to the right of the peak. This explanation, however, fails to explain the hint of a harmonic peak near $k = 0.04$ Å^{-1} in figure 5.

Finally, Hohr *et al.* (1988) suggest that the shape of the curve in the intermediate régime is because of strongly interacting particulate domains. They use liquid theory to calculate an approximate structure factor which shows harmonic peaks. Although Hohr *et al.* (1988) believe structure develops in the bulk

FIGURE 11. Transmission electron micrograph of acid leached 10–30–60. Sample was leached at pH = 1 for two days at 70 °C. Scale bar represents 0.1 µm.

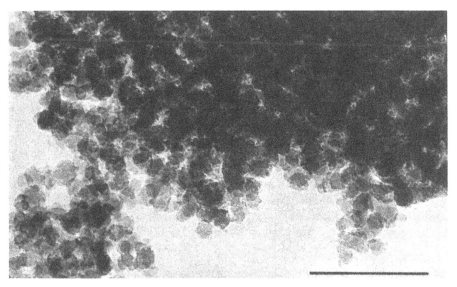

FIGURE 12. Transmission electron micrograph of 10–30–60 leached at pH = 7 for seven days at 70 °C. Scale bar represents 0.1 µm.

glass, their calculation could equally apply to leach-induced nucleation. This analysis is consistent with the position adopted by us above, that the observed structure is created by leach-induced nucleation and growth.

To test the above ideas, we leached 10–30–60 under different conditions and demonstrated that the scattering profile definitely depends on the leach protocol. Figure 10 compares acid- and distilled-water (DW) leached 10–30–60. Neutral leaching leads to fractally rough surfaces on short scales ($P = -3.5$), whereas distinct surfaces are absent in the acid-leached case. Apparently acid leaching drastically disrupts the silica network (see figure 11 for an electron micrograph that shows the absence of surfaces for acid leaching and see figure 12, which shows rather distinct particulates under neutral leaching).

The peak in the scattering curve also depends on the leach conditions. If a pre-existing domain structure determined the leach pattern, the peak should remain fixed, and independent of leachate. We conclude that the structure of leached borosilicates is more dependent on the leaching process than on phase separation in the bulk glass.

4. MICELLAR SOLUTIONS

Micellar solutions are another system in which kinetic processes might be observed in the early stages of phase separation. Micellar solutions have two features conducive to the observation of nonequilibrium structures. First, the basic structural unit is a 50 Å cluster of surfactant molecules. In a sense, a micellar solution is like a simple fluid with 50 Å molecules. We expect long timescales for structural rearrangement, enhancing our ability to observe non-equilibrium domains. In addition, because surface tension between the phases is exceeding small, the driving force for the formation of compact domains is reduced.

We choose a solution of non-ionic surfactant, n-dodecylhexaoxyethylene glycol monoether ($C_{12}E_6$), in D_2O. The critical point in this system is a lower consolute point at 48.17 ± 0.03 °C and 2.5 ± 0.5% surfactant (by mass). The system phase separates on raising the temperature. Quench experiments are done by holding the system about 0.05 °C below the critical temperature and then quenching by raising the temperature slightly above the phase boundary. Temperature equilibrium is reached in 3 min. We collect time slices of SANS and light scattering data in intervals of 10 min. The shape of the scattering curves changes little over 2 h (Wilcoxon *et al.* 1988).

Figure 13 compares the scattering for a critical and off critical quench to the same temperature increment ϵ (such that $\epsilon = (T_c - T/T_c)$, where T_c is the consolute temperature). On the critical isochore, we observe power-law correlations over nearly three decades in length scale with $D = -P = 1.6$. These fractal fluctuations persist for many hours but eventually disappear indicating approach to equilibrium. Note that for this system, the spinodal peak occurs at $k < 0.0001$ Å^{-1} (Kuwahara *et al.* 1985). Although a spinodal ring is present we do not detect it in the k régime under study.

Fractal domains are not observed when an off-critical mixture is quenched to an equivalent depth. Instead, the observed profile corresponds to weakly interacting collection of micelles.

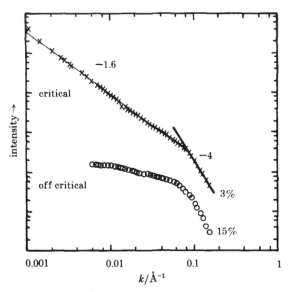

FIGURE 13. Scattering data for phase separating $C_{12}E_6$ micellar solutions; $\epsilon = -6.2 \times 10^{-4}$. Fractal correlations are observed only for the critical quench. The break in the critical quench data occurs at k^{-1} approximately equal to the micelle radius, indicating that the primary aggregates remain intact during phase separation. The curves change little over a two hour period following the quench (from Wilcoxon et al. 1988). Percentages are by mass.

The high-k region of both data sets in figure 12 decay with a power-law exponent ca. -4. The shape in this region is because of the form factor of the 50 Å micelles. This form factor changes little during the quench indicating that the basic micellar structure remains intact.

Although we do observe fractal patterns, our data do not confirm the simulations of Grest & Srolovitz (1984) or those of Desai & Denton (1986). These authors observed fractal structures that are strongly overlapped. Because of this overlap, fractal scattering behaviour is unobservable on large length scales. Furthermore, the absence of ramified patterns in the off-critical quench suggests that non-compact nucleation is not observable in this system. Our data therefore do not support a kinetic growth mechanism for the generation of fractal domains.

Based on the above discussion and further observations on the dynamics of concentration fluctuations in the two-phase régime (Wilcoxon 1988), we believe that the fractal correlations are long-lived remnants of intermicellar correlations that develop in the single phase regime very near the critical point. To test this idea we studied the single-phase régime very close to the critical temperature. Figure 14 shows temperature dependence of the scattering profile on the critical isochore in the single-phase régime.

Figure 14 shows that as the critical point is approached, strong deviations form O.Z. behaviour are observed. A large region of k space opens up where fractal correlations are observed with a Porod slope $P = -1.4$, substantially different

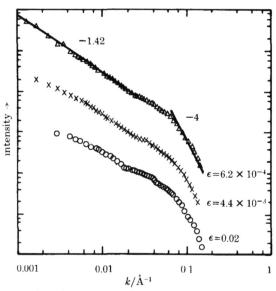

Figure 14. Scattering data for micellar solutions in the single-phase region on the critical isochore as a function of temperature. Near the critical point ($\epsilon = 0$), fractal correlations are observed on length scales between 50 and 1000 Å. The micelle form factor is evident at large k

from the O.Z. predication of -2.0. At very small k (in the light-scattering régime) the data do follow the O.Z. behaviour. It appears then that this system shows unusual fractal correlations on length scales larger than the micelle size but small compared to the O.Z. correlation range. As the critical point is approached, the correlation range is so large that the fractal correlations cover almost two decades in length scale.

All of our observations are consistent with the growth of polymeric micelles at equilibrium in the single phase region. At a given temperature, flexible chains form whose length increases with both concentration and temperature. Because the crossover at large k is preserved as fractal correlations develop, either the original quasispherical micelles remain intact, or tubular structures are formed whose diameter is that of the parent micelles. The latter seems more reasonable.

The observed fractal dimensions of 1.4–1.6 are somewhat below that expected for both ideal ($D = 2$) and self-avoiding linear chains ($D = 1.7$). Both the concavity of the scattering data below $k = 0.01$ Å^{-1} and the low value of D are consistent with local chain stiffness. For stiff chains a crossover toward $D = 1$ is expected on length scales short compared to the persistence length.

Within the polymer interpretation, the observed k^3 dependence of the linewidth in the two-phase region (Wilcoxon *et al.* 1988) is attributable to internal relaxation processes arising from the Rouse–Zimm modes of the chain. In contrast to conventional polymers, however, correlation functions are exponential. The difference is probably caused by the fact that relaxation process are driven by elastic rather than osmotic forces (Safran *et al.* 1984).

From thermodynamic arguments (Blankenstein *et al.* 1986) near-equilibrium polymerization is favoured with increasing concentration and increasing temperature. Along the critical isochore, micelles of increasing molecular mass develop and phase separation occurs when the molecular mass is high enough that neighbouring chains just overlap (c^* in the polymer language) analogous to polymers near the θ-point. At T_c, phase separation again occurs with increasing concentration when the molecular mass is sufficient that chains just overlap.

At concentrations above the critical isochore, polymeric micelles overlap and entangle as in semidilute polymer solutions (Candau *et al.* 1985). Fractal correlations become unobservable because of chain overlap. That is, as the system is uniform on scales large compared to ξ, fractal correlations can only be observed when ξ is large compared to the monomer size, the diameter of the parent micelles. This condition is only achieved near or below the critical temperature.

In the two-phase region, the slow approach to equilibrium suggests that the polymeric structure is preserved. At equilibrium, the rich phase is like a semidilute solution of entangled polymeric micelles whereas the poor phase is a dilute solution of collapsed polymeric micelles. Measured hydrodynamic radii in the two phases are consistent with this interpretation.

In summary, all of the observed properties of micellar solutions near the coexistence curve are consistent with the growth of stiff polymer chains whose diameter is that of the parent micelles. It should be re-emphasized that this discussion is concerned with short-length-scale structure and that ordinary critical behaviour is expected and observed on longer scales (Wilcoxon *et al.* 1989).

5. CONCLUSION

Fractal geometry provides a new prospective from which to analyse patterns that appear in phase-separated systems. Although several theories and simulations of phase separation predict transient fractal patterns, the patterns arising in the systems studied here seem to have their origin in unexpected phenomena. For leached glasses, it is the leaching process itself that is of dominant importance. The 'spinodal-like' peak in the structure factor arises during leaching. This peak, however, appears to be caused by a nucleation and growth process induced by leaching. The short-range fractal surfaces also result from leaching via reaction-limited dissolution. For the micelle system, fractal correlations are found in the single-phase régime where kinetic growth processes seem unlikely. Here the growth of stiff polymer chains near equilibrium in the single-phase region provides a more reasonable interpretation. Although the original motivation of our work was to seek early-stage kinetic growth processes, we are unable to trace any of our observations to these processes.

We thank T. J. Headley for the electron microscopy reported here.

REFERENCES

Berk, N. F. 1987 *Phys. Rev. Lett.* **58**, 2718–2722.
Blankschtein, D., Thurston, G. M. & Benedek, G. B. 1986 *J. chem. Phys.* **84**, 955–958.

Bunker, B. C., Headley, T. J. & Douglas, S C. 1984 In *Better ceramics through chemistry* (ed. C. J. Brinker, D E. Clark & D. R. Ulrich) (*Mat. Res. Soc Symp. Proc., Pittsburgh*), vol 32, pp. 41–46.
Bunker, B. C., Arnold, G W., Day, D. E & Bray, P J. 1986 *J. non-cryst Solids* **87**, 226–253
Cahn, J. W 1965 *J chem. Phys* **42**, 93–99.
Candau, S. J., Hirsch, E & Zana, R. 1985 *J. Colloid Interface Sci.* **105**, 521–528.
Desai, R C. & Denton, A R. 1986 In *On growth and form* (ed. H E. Stanley & N Ostrowsky, pp. 237–243. Dordrecht: Martinus Nijhoff.
Grest, G. S. & Srolovitz, D. J 1984 *Phys Rev.* B **30**, 5150–5155.
Guenoun, P., Perrot, F & Beysens, D 1987 *Phys. Rev* A **36**, 4876–4890.
Haller, W. & Macedo, P. B 1968 *Physics Chem. Glasses* **9**, 153–155.
Hohr, A , Neumann, H -B , Schmidt, P W., Pfeifer, P. & Avnir, D. 1988 *Phys. Rev.* B **38**, 1462–1467.
Hurd, A. J., Schaefer, D. W. & Martin, J. E 1987 *Phys. Rev.* A **35**, 2361–2364.
Kuwahara, N., Hamano, K. & Koyama, T. 1985 *Phys. Rev.* A **32**, 1279–1281
Mandelbrot, B. B. 1982 In *Fractal geometry of nature*. San Francisco: Freeman.
Safran, S. A., Turkevich, L A. & Pincus, P. 1984 *J Phys. Lett* **45**, L69–L74.
Schaefer, D. W., Bunker, B. C. & Wilcoxon, J. P. 1987 *Phys Rev. Lett.* **58**, 284–285.
Schaefer, D. W. 1988 *MRS Bulletin* vol. XIII (2), 22–27.
Schaefer, D. W , Hurd, A J & Glines, A. M 1988 In *Fluctuations and pattern growth: experiments and theory* (ed. H. E. Stanley and N. Ostrowsky) Dordrecht Kluwer Academic Publishers.
Schmidt, P W., Steiner, M., Hohr. A., Neumann, H.-B & Avnir, D. 1987 In *Fractal aspects of materials disordered systems* (ed A. J. Hurd, D. A. Weitz & B. B Mandelbrot) (*Mat. Res Soc Symp. Proc.. Pittsburgh*), vol. EA-13, pp. 121–124.
Sinha, S. K., Drake, J M., Levitz, P. & Stanley, H. E. 1987 In *Fractal aspects of materials disordered systems* (ed. A. J. Hurd, D. A. Weitz & B. B. Mandelbrot (*Mat. Res Soc. Symp. Pittsburgh*), vol. EA-13, pp. 118–120.
Wilcoxon, J. P., Schaefer, D. W & Kaler, E. W. 1988 *Phys. Rev Lett.* **60**, 333–336.
Wilcoxon, J P , Schaefer, D. W. & Kaler, E. W. 1989 *J. chem Phys.* (In the press.)
Wiltzius, P., Bates, F. S , Dierker, S. B & Wignall, G. D. 1987 *Phys. Rev* A **36**, 2991–2994

Discussion

J. S. ROWLINSON (*Physical Chemistry Laboratory, University of Oxford, U.K.*). Dr Schaefer has shown that in micellar solutions one can see clearly the effects of the critical point singularity (in, for example, the susceptibility) for up to 30 K into the one-phase region, and he has interpreted this behaviour in terms of polymer entanglements. I would caution against so specific an interpretation, because such traces of critical behaviour are found also in the one-phase region of much simpler systems. Thus for argon at its gas–liquid critical point we can use the configurational part of the heat capacity at constant pressure as a measure of the susceptibility. This quantity is infinite at the critical point itself and shows strong maxima on isotherms 20 and 30 K above the critical temperature of 150 K. Thus even at 180 K the pseudo-critical peak in C_p accounts for over half of the observed value. This temperature is 30 K above T_c, and would correspond to a difference of about 70 K in systems such as those considered by Dr Schaefer for which T_c is about 350 K.

D. W. SCHAEFER. Professor Rowlinson's warnings are certainly valid. We issue the same caution elsewhere (Wilcoxon *et al.* 1989). Indeed, we have observed normal Ising-like critical phenomena in the light-scattering régime on these systems (Wilcoxon & Kaler 1987; Wilcoxon *et al.* 1989).

There are differences between micellar systems and simple fluids: the coexistence curve is distorted strongly toward low concentrations and its general shape is characteristic of polymers. In addition, at intermediate length scales, we observe the fractal exponents which are not consistent with critical fluctuations as predicted by Ornstein–Zernike theory. To our knowledge, such deviations have not been observed in simple fluids. These observations can be explained, however, if the critical system is polymer-like on intermediate length scales. On large length scales, normal critical phenomena is expected and observed. On intermediate length scales the unusual fractal behavior we have observed is consistent with chain-like structure.

Our observations may be the structural consequences of the micellar growth model of Blankenstein et al. (1986). These authors postulate the growth of rod-like micelles and adequately account for both the shape of the coexistence curve and the susceptibility in the single-phase region. Our work indicates that the rods are better described as polymers and that they persist into the two-phase region.

Reference

Wilcoxon, J. P. & Kaler, E. W. 1987 *J. chem. Phys* **86**, 4684.

R. C. BALL (*Cavendish Laboratory, University of Cambridge, U.K.*). Are the subcritical fluctuations to which Dr Schaefer applies a 'polymeric' interpretation experimentally distinct from the critical fluctuations in his micellar system?

D. W. SCHAEFER. In this paper we concentrated primarily on fluctuations between 10 and 500 Å as measured by neutron scattering. In this régime, the fluctuations are definitely distinct from critical fluctuations that we have measured by light scattering (Wilcoxon & Kaler 1987; Wilcoxon et al. 1989). On longer length scales, normal Ornstein–Zernike theory applies and we observe power-law scattering profiles with slopes of -2. Critical exponents as measured by light scattering are also consistent with ordinary critical phenomena on long length scales. The unusual behavior occurs at shorter length scales (less than 1000 Å) and appears to be yielding information about the structure of the system that is going critical.

The fact that fractal correlations persist into the two-phases régime for several hours suggests that these fluctuations are not caused by intermicellar correlations, which should dissipate rapidly. Rather they indicate the structure of the micelle itself.

D. J. TILDESLEY (*Department of Chemistry, University of Southampton, U.K.*). Would Dr Schaefer comment on the nature of the two coexisting phases above the cloud point? Does he see remnants of the 70 Å micelle polymers, observed during the nucleation in the high-density phase at equilibrium?

D. W. SCHAEFER. If this system really is polymeric, then the nature of the two existing phases should be very similar to that observed in ordinary polymers. The rich phase is a semidilute polymer solution and the pore phase consist of collapsed polymer chains.

Because of time limitations, we did not measure the structure of the high-density phase at equilibrium. We know that the remnants of the 70 Å micelles persist for at least 2 h but cannot definitively say they exist at equilibrium. Our opinion, however, is that they do persist. Equilibrium dynamics (Wilcoxon *et al.* 1989) are consistent with this viewpoint.

Experiments on the structure and vibrations of fractal solids

By E. Courtens[1] and R. Vacher[2]

[1] *IBM Research Division, Zürich Research Laboratory, 8803 Rüschlikon, Switzerland*
[2] *Laboratoire de Science des Matériaux Vitreux, Unité Associée au CNRS, No. 1119, Université des Sciences et Techniques du Languedoc, F-34060 Montpellier, France*

Measurements performed on carefully prepared series of silica aerogels are reported. These materials were studied with small-angle neutron scattering, revealing a fractal structure that can extend up to at least two orders of magnitude in length.

The corresponding collective vibrations were investigated by Brillouin, Raman, and inelastic neutron scattering. All results are consistently explained in terms of fractons.

Introduction

Aerogels are monolithic solid materials with an extremely tenuous microscopic structure (Fricke 1985; Vacher *et al.* 1989*a*). The most thoroughly investigated aerogels are made of silica. These can be prepared with a porosity as high as 99%. In consequence, they exhibit unusual physical properties, making them suitable for a number of technical applications, such as Cerenkov radiators, supports for catalysts, or thermal and acoustic insulators. Suitably prepared aerogels are also excellent examples of fractal solids (Vacher *et al.* 1988*a*). Thus it is both of fundamental and technical interest to determine the structure of aerogels, to investigate the mechanisms of their formation, and to study their vibrational dynamics.

The starting point for the preparation of silica aerogels is the hydrolysis of an alkoxysilane $Si(OR)_4$, where R is usually CH_3 or C_2H_5 (Kistler 1932). The reaction is strongly influenced by the amount of 'catalyst', either acid or base, that can be added to the water. Important other parameters are the relative concentrations of reagents, usually expressed by the ratio $[Si(OR)_4]/[H_2O]$, and the amount of alcohol ROH used to dilute the reagents. The ratio $[Si(OR)_4]/[ROH]$ determines the final density of the aerogel. Hydrolysis produces —SiOH groups that polycondense into —Si—O—Si— bonds. Small particles start to grow in the solution. These particles bind solidly, and eventually form a disordered network filling the reaction vessel. At this point the solution gels. The above mechanisms can of course occur simultaneously rather than sequentially. After gelation, the so-called 'alcogel' continues to evolve in time because the reactions are generally not complete at the gel point. To obtain the solid porous structure, the solvent is finally removed after some 'ageing' time. If this is done simply by evaporation in air, the capillary forces at the liquid–vapour interfaces pull on the silica network, causing considerable internal fracture and resulting in macroscopic shrinkage. This leads to so-called 'xerogels' of modest porosities (Fricke 1985). However, if the solvent is extracted above its critical point, the microscopic structure of the network is preserved, and extremely porous 'aerogels' can be obtained (Kistler 1932). To perform this hypercritical extraction, the alcogel and additional solvent are introduced in a bomb, the temperature and pressure are raised, and the vapour is allowed to escape with the temperature maintained above the critical point. By this technique, large blocks of silica aerogels can be prepared whose prominent physical properties are lightness, optical translucency, solid-like elasticity, and very high internal surface.

The gel formation is clearly controlled by the physical and chemical mechanisms of reaction and aggregation. For silica, there is a complex set of reactions, strongly influenced by the ionicity (Iler 1979). To our knowledge, there has so far been no computer simulation directly applicable to this case. However, a number of relevant models has been studied in great detail. They have shown that in general fractal networks are formed. Some limiting behaviours for growth in three-dimensional euclidean space are listed in table 1. Reaction-limited cluster–monomer aggregation (Eden model) leads to a compact object (Hausdorff dimension $D = 3$) whose surface is fractal (Eden 1961; Meakin 1983a). Diffusion-limited cluster–monomer processes (Witten–Sander model) lead to $D = 2.5$ (Witten & Sander 1981; Meakin, this symposium). At the other extreme, the diffusion-limited aggregation of clusters leads to very tenuous structures with $D = 1.78$ (Meakin 1983b; Kolb et al. 1983).

TABLE 1. FRACTAL DIMENSION OF AGGREGATES

| | aggregation limited by | |
aggregation mechanism	reaction	diffusion
cluster–monomer	3	2.50
cluster–cluster	2.1	1.78

The results of our small-angle neutron scattering (SANS) presented below demonstrate that aerogels that are fractal over more than two orders of magnitude in length scale can be prepared. Depending on catalysis conditions, D varies between ca. 1.8 and ca. 2.4. Furthermore, the elementary particles can be smooth or rough, and their size strongly depends on catalysis. The variation of the preparation parameters is obviously sufficient to create vastly different aggregation conditions. In particular, for base-catalysed aerogels, it appears that diffusion-limited cluster–cluster aggregation may play a major role in the final formation of the gel (Schaefer et al. 1988).

The exceptional thermal and acoustic properties of silica aerogels originate from the hindrance to propagation of acoustic vibrations in their extremely porous structure. As SANS has unambiguously demonstrated a fractal geometry in these materials, the concept of 'fractons', the elementary vibrational excitations of fractal elastic networks, should be relevant (Alexander & Orbach 1982). Theory predicts that fractons are strongly localized excitations, whose characteristic size l is related to the frequency ω by a 'dispersion law', $\omega \propto l^{-D/\bar{d}}$, where $\bar{d}$ is the so-called spectral or fracton dimension (Orbach 1985 and references therein) A related theoretical prediction is that the vibrational density of states (DOS), $Z(\omega)$, is proportional to $\omega^{\bar{d}-1}$. One expects $\bar{d}$ to be much smaller than D (Alexander & Orbach 1982). Thus, fractons are very different from acoustic phonons. We have been able to confirm some of these predictions in silica aerogels, and in particular to determine the exponents D and $\bar{d}$.

FRACTAL STRUCTURE OF AEROGELS

For a mass fractal, the mass M scales with the size L as $M \propto L^D$, with $1 \leqslant D \leqslant 3$. In a real material, if one assumes that fractal clusters extend up to a correlation length ξ, the density–density correlation function $g(r)$ can be modelled by

$$g(r) - 1 \propto r^{D-3} \exp(-r/\xi). \tag{1}$$

The exponential decay is introduced to account for the finite extent of fractal clusters. The space Fourier transform of equation (1) is proportional to the structure factor $S(q)$ and to the scattered intensity (Teixeira 1986). The Fourier variable q is also the momentum exchange in scattering. Two limiting régimes are of interest. At small q, $q\xi \ll 1$, a pseudo-Ornstein–Zernicke behaviour is observed, $S(q) \propto 1/(1+q^2\xi^2)^{\frac{1}{2}(D-1)}$. On the other hand, when $q\xi \gg 1$, $S(q)$ is nearly proportional to q^{-D}. At even larger q, such that the neutron wavelength is smaller than the average particle radius R, scattering originates from the particle surfaces. Fractal surfaces are characterized by a surface fractal dimension, $2 \leqslant D_s \leqslant 3$. In this régime, $S(q)$ is approximately proportional to q^{D_s-6} (Bale & Schmidt 1984; Wong & Bray 1988). For smooth particles, $D_s = 2$, one recovers the usual Porod law, $S(q) \propto q^{-4}$ (Guinier 1956). Finally, at very large q, comparable to the inverse of the bond size, the atomic arrangement determines the scattering. In principle, the measurement of $S(q)$ on a fractal can give the four parameters ξ, R, D, and D_s that characterize the structure. It is the quality of the fit of the experimental data to the Fourier transform of equation (1) that justifies the phenomenological exponential cut-off.

The SANS experiments were done on the spectrometer PACE, at the Laboratoire Léon Brillouin† in Saclay, France. In this system, the wavelength of the incident neutrons can be varied from *ca.* 4 Å to *ca.* 22 Å‡. Elastically scattered neutrons are collected on a set of 30 annular detectors, whose radii range from 3 to 33 cm. The distance between sample and detectors can be varied from *ca.* 1 m to *ca.* 5 m, allowing a range of q from 2×10^{-3} Å^{-1} to 0.5 Å^{-1}.

Two series of samples were used, both prepared from tetramethoxysilane dissolved in methanol. In each series, different samples correpond to different densities. The first series was obtained under neutral conditions, whereas the second was prepared under basic catalysis. The upper three curves of figure 1 are examples of spectra obtained on samples prepared under neutral conditions. The most striking feature of these curves is the power law observed for the

---

† Laboratoire Léon Brillouin is a joint laboratory Centre National de la Recherche Scientifique–Commissariat à l'Energie Atomique.

‡ 1 Å = 10^{-10} m = 10^{-1} nm.

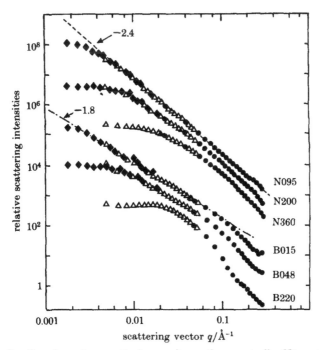

FIGURE 1. Small-angle neutron-scattering results on three neutrally (N) reacted samples (top) and three base (B)-catalysed samples (bottom). The numbers in the labels refer to the density in kilograms per cubic metre. The straight lines are guides to the eye, indicating the large fractal range of light samples, and the effect of catalysis on the Hausdorff dimension D. The different symbols refer to different runs (Vacher *et al.* 1988a).

lightest sample over about two orders of magnitude in q. A fractal dimension $D = 2.40 \pm 0.03$ is obtained from the fit of each individual curve to the theoretical expression for $S(q)$. For the lightest samples, fractal geometry seems to extend down to the smallest length scale probed in this experiment. For aerogels with densities larger than 200 kg m^{-3}, the departure of $S(q)$ from the $q^{-2.4}$ dependence at large q indicates the presence of particles with gyration radii of a few ångströms. An expanded plot of the large-q range for the heaviest samples shows a region where $S(q)$ is nearly proportional to q^{-3} above *ca.* 0.15 Å^{-1}. This is best seen in a presentation of the scattered intensity times q^3, on figure 2. The plateau indicates fuzzy particles with a fractal surface. The structure at that scale can be modified by oxidation at 500 °C, a treatment that removes remaining —CH$_3$ groups and creates new siloxane bonds. After such a treatment, we observe $S(q) \times q^3 \propto q^{-1}$ at large q, as shown in figure 2. This demonstrates that oxidation smoothens the surface of the particles.

From the upper three curves in figure 1 one also sees that ξ strongly decreases with increasing density. Using the values measured on the whole series of samples, one obtains $\rho \propto \xi^{-0.60 \pm 0.02}$ (Vacher *et al.* 1988a). To interpret this scaling law, let us consider elementary fractal clusters assembled from homogeneous particles of size a. In three-dimensional euclidean space, the density of such clusters at a

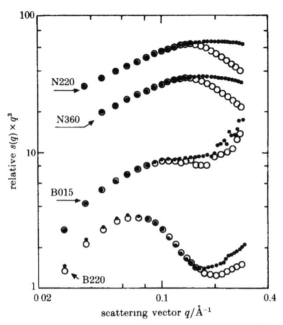

FIGURE 2. Small-angle neutron-scattering intensities multiplied by q^3 for two neutrally reacted and two base-catalysed samples. Only the runs at highest q-values are shown to emphasize the particle region The full circles refer to unoxidized samples and the open ones to oxidized samples. The labelling is the same as in figure 1.

length scale $L > a$ is $\rho(L) = \rho_a(L/a)^{D-3}$. Here, ρ_a is the density of the particles. Let ξ be the final average size of the clusters, and assume that the aerogel is built of an homogeneous assembly of such clusters. Hence, the bulk density of the aerogel is $\rho \equiv \rho(\xi)$. If we now construct another aerogel in exactly the same way, only changing the value of ξ, we obtain two materials of different bulk density, but which are identical at length scales at which they are fractal. We have named this property 'mutual self-similarity' (Vacher et al. 1988a). For these materials, the scaling law obeyed by ρ as a function of ξ *on the whole series of samples* is identical to the relation between $\rho(L)$ and L in the fractal region for *any one* sample. For our aerogels, $D = 2.4$ implies $\rho(L) \propto L^{-0.6}$ for each sample. Because an identical exponent is found for $\rho(\xi)$, the mutual self-similarity of our series of samples prepared under neutral conditions is established. Hence, a study of the dependence on ρ of macroscopic properties on such a series amounts to studying the scaling of those properties as functions of the length on a single fractal.

The lower three curves in figure 1 are examples of $S(q)$ for samples prepared under basic catalysis (Vacher et al. 1988b). They are very different from the upper three curves. First, they show that basic catalysis leads to the formation of *large* particles, with $10 \text{ Å} \leqslant R \leqslant 20 \text{ Å}$ weakly dependent on the aerogel density. Further, comparing N200 to B220, one notes that the extension of the power-law region is very different on these two curves. For this density, corresponding to a porosity of *ca.* 90%, the sample prepared under basic conditions does not exhibit a clear fractal behaviour (Schaefer & Keefer 1986). To obtain in this case a fractal structure over two orders of magnitude, it is necessary to prepare materials of extremely low densities, as illustrated by curve B015. For the latter, the power-law behaviour extends down to the smallest q values accessible in this SANS experiment. Finally, a most striking effect of catalysis is found in the value of D. All curves from base-catalysed samples with an extended fractal range give $D = 1.8 \pm 0.1$.

For base-catalysed samples, the roughness of the particles is function of the density, as illustrated by the lower two curves of figure 2. For $q \approx 0.1$ Å^{-1}, the composition B015 exhibits a plateau suggesting $D_s \approx 3$, whereas for B220 the particles are smooth. The increase in $S(q) \times q^3$ above $q \approx 0.2$ Å^{-1} for these curves probably originates in the tail of the structure factor peak of silica, expected at $q \approx 1.5$ Å^{-1}. It is worth noting that the wing of that peak contributes to the B samples and not to the N samples. One should also notice that oxidation never has an effect for $q \lesssim 0.15$ Å^{-1}. In particular it does not change the surface fractality of large particles, whereas it completely smoothens the particles of neutrally reacted samples. These facts emphasize that oxidation has an effect only at the size of the attached —CH$_3$ groups.

ACOUSTICAL VIBRATIONS OF FRACTAL AEROGELS

The vibrational régimes

As seen in the above discussion of the structure, one recognizes three distinct length regions in a fractal aerogel. Below the particle size a the material is compact with density ρ_a, in the intermediate region it is a mass fractal of Hausdorff dimension D, and above the correlation length ξ it becomes a light homogeneous material of density $\rho \ll \rho_a$. Related to these three structural regions, one expects three distinct vibration régimes, as illustrated in figure 3. Corresponding to the macroscopic homogeneous region, one anticipates acoustic phonons. At very low frequencies, these will be damped by relaxation processes, giving a linewidth Γ approximately proportional to ω (Michard & Zarembowitch 1988). At higher frequencies, elastic (Rayleigh) scattering can become dominant, giving $\Gamma \propto \omega^4$, as actually observed in Brillouin experiments (Courtens et al. 1987). Corresponding

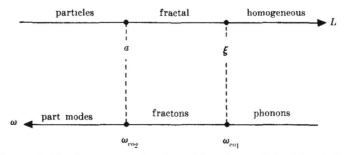

FIGURE 3. The three structural regions of fractal aerogels (top) in relation to their vibrational régimes (bottom).

to the fractal region, fractons are anticipated with an elastic linewidth $\Gamma \propto \omega$ (Aharony *et al.* 1987), and with the unusual dispersion and DOS indicated in the Introduction. Finally, to the particles correspond vibration modes, whose DOS increases with ω for the surface contributions, and with ω^2 for the bulk contributions.

In figure 3, the transitions between these different régimes are indicated by simple crossovers at ω_{co1} and ω_{co2}. The details of these crossovers can be considerably more involved. In particular, a single ω_{co1} should apply to the phonon–fracton crossover only if the frequency of phonons of wavelength $2\pi\xi$ equals that of fractons of characteristic size ξ, and if at the same frequency the Ioffe–Regel limit $\Gamma = \omega$ is just reached. Otherwise, successive crossovers could occur (Aharony *et al.* 1987). In the following, a single ω_{co1} is assumed. However, the smooth crossover experimentally observed (Courtens *et al.* 1988) could be an indication for a succession of crossovers that cannot be extracted reliably from such measurements. Similarly, one could conceive that the frequency of fractons of characteristic size a were different from the lowest vibrational eigenmodes of the particles. For the neutrally reacted aerogels, we have found experimentally that those frequencies are in fact similar (Tsujimi *et al.* 1988).

The phonon–fracton crossover

Neutrally reacted aerogels can be prepared with $2\pi\xi \sim 0.3\,\mu\text{m}$. Thus the crossover region can be investigated with Brillouin scattering of visible light. Indeed, that technique detects vibrational excitations of wavelengths down to $\lambda/2n$, where λ is the optical wavelength and n the refractive index of the material. Polarized scattering was measured for several values of the momentum exchange q, and for a series of samples of different ρ, as explained in detail elsewhere (Courtens *et al.* 1987, 1988). It turns out that, for neutrally reacted aerogels, the frequency position of the peak in the scattered spectra falls into the measurement range of Brillouin interferometry. At sufficiently small q, or sufficiently large ρ values, phonons are observed with a nearly lorentzian spectral profile. These lines broaden rapidly with increase of q. As the crossover region is reached, the lineshapes become strongly non-lorentzian, and the position of the peak becomes stationary with further increase in q.

The experimental lines could be fitted with a theoretical profile derived by using an effective medium approximation (Polatsek & Entin-Wohlman 1988). The fits used heuristic expressions for the frequency dependence of the elastic lifetime and of the dispersion law near crossover. A non-critical parameter m fixes the sharpness of this crossover (Courtens *et al.* 1988). The value $m = 2$ was found satisfactory for all our measured spectra. The fits depend then only on three relevant parameters: the crossover frequency ω_{co1}, the associated wavevector q_{co}, and the overall intensity. Figure 4 illustrates the quality of such fits. It also demonstrates that a sharp crossover (m very large) is inadequate, a feature to be kept in mind in the interpretation of other spectroscopic data.

The samples of this series being mutually self-similar, the points ω_{co1} against q_{co} for various materials must all fall on the same fracton 'dispersion curve'. This is shown in figure 5, where the different symbols correspond to different samples,

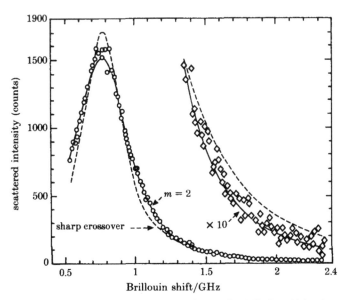

FIGURE 4. Typical fits to a Brillouin spectrum (N200, $\theta = 90°$) for which q is near q_{co}. The solid line is for $m = 2$, whereas the broken line is the best fit obtained with a very large value of m (sharp crossover). Clearly, the smooth crossover gives a superior fit over the entire frequency range.

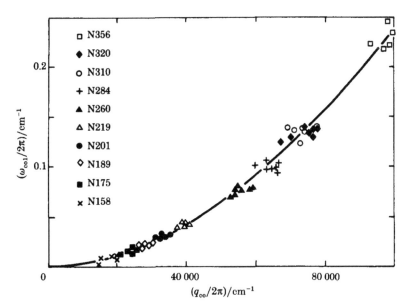

FIGURE 5. The fracton dispersion curve derived from the Brillouin determination of ω_{col} and q_{co} on a series of mutually self-similar samples.

whereas the same symbol is used for different q-values in the measurements. One notes that all points cluster on the curve $\omega \propto q^{D/\bar{d}}$, covering more than one order of magnitude in ω. This is the first fracton 'dispersion curve' to be determined experimentally. From the fit one extracts $D_{ac}/\bar{d} \approx 1.9$. Here, D_{ac} is an 'acoustical' fractal dimension to be related to the 'connectivity' rather than to the mass.

The crossover wavevector q_{co} equals $1/\xi_{ac}$, where ξ_{ac} is an acoustical correlation length. To the extent that only one length scale is relevant to determine the upper limit of fractality, one anticipates $\xi_{ac} \propto \xi$, the latter being determined by SANS. The two quantities are shown in figure 6. As explained above, the slopes are given by $1/(D-3)$. From the Brillouin data, the 'acoustical' value is $D_{ac} = 2.46 \pm 0.03$, in satisfactory agreement with D from SANS. One also notes that ξ_{ac} is approximately five times larger than ξ. This numerical factor could be mostly caused by the different definitions that are used for the two quantities. Whereas in the SANS measurements the fractality extends from the gyration radius R to ξ, in the acoustical measurement it covers the range from a to ξ_{ac}, where a is the average particle diameter, $a \approx 2\sqrt{\tfrac{5}{3}}R$. This can account for a factor of ca. 2.5. Another factor could be because of the sensitivity of the acoustic measurement on connectivity, whereas SANS senses the mass. Fractality in the connectivity can extend further than mass fractality.

From $D_{ac}/\bar{d}$ and D_{ac}, and taking into account the uncertainty in m, one finds $\bar{d} = 1.3 \pm 0.1$. This value happens to be close to the 'scalar elastic' (Alexander 1984) prediction $\bar{d} = \tfrac{4}{3}$ (Alexander & Orbach 1982). It is rather different from the

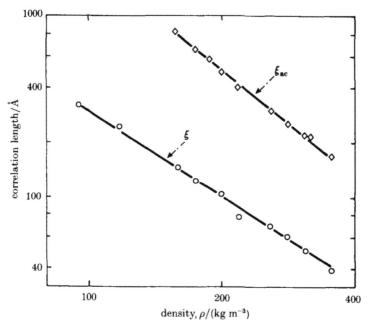

FIGURE 6. The acoustically determined correlation length ξ_{ac} compared with the SANS value ξ, for the mutually self-similar series of neutrally reacted aerogels.

tensorial elastic result calculated on percolation clusters in three dimensions, $\bar{d} = 0.9$ (Webman & Grest 1985). This should probably be taken as an indication that the aerogels have a more rigid structure than percolation clusters, rather than as an evidence for scalar elasticity, as discussed in the Conclusions.

Light scattering from fractons

The effective medium light-scattering formula predicts that the asymptotic behaviour of the scattered intensity at high ω, in the fracton régime, should be $I \propto \omega^{-3-2\bar{d}/D}$ (Courtens et al. 1988). This is obeyed fairly well for samples that are not too light, or for ω values that are not too large, as illustrated in figure 7. However, for the lightest samples in figure 7, one notices that the effective slope becomes smaller than 4 in absolute value. This is presumably because of a weakness of the theory, which does not describe adequately the scattering from fractons when these have a spatial extent much smaller than the optical wavelength. In that case, each fracton scatters independently, and the scattered intensity is the sum of intensities, rather than the square of the sum of amplitudes as postulated by the Green function formalism. Because the individual fractons have fairly arbitrary spatial shapes (Yakubo & Nakayama 1987), they are

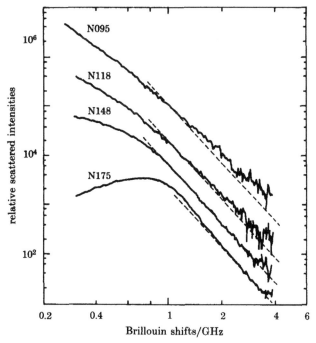

FIGURE 7. Logarithmic presentation of the polarized scattered intensity in the backward direction for four light samples of the neutrally reacted series. The dark counts of the photomultiplier have been subtracted. The lines have a slope $-3-2\bar{d}/D_{ac}$, and are guides to the eye. Whereas for the two densest samples there is a region at $\omega > \omega_{col}$ that appears to agree with that slope, an obvious additional component, growing with ω, contributes to the scattered intensities of the two other samples.

expected to depolarize the light quite effectively, and indeed the additional scattering is found to be strongly depolarized.

We have measured this scattering in the backward geometry, and in VH polarization† (Tsujimi *et al.* 1988). In such conditions there is no contribution from the phonons. The results can be plotted as $I(\omega)/n(\omega)$ against ω. Here, $n(\omega)$ is the Bose–Einstein factor representing the thermal population of the fractons. One finds a smooth crossover into the fractons, followed by a power law I/n approximately proportional to $\omega^{-0.4}$, followed near ω_{co2} by particle modes. The slope in the fracton region can be expressed by (Tsujimi *et al.* 1988)

$$I/n(\omega) \propto \omega^{-2+2d_\phi \bar{d}/D}. \tag{2}$$

Here d_ϕ is an exponent relating the strain ϵ to the mean euclidean dimension l of the fracton, $\epsilon \propto l^{-d_\phi}$.

Based on current literature, this was originally interpreted as a superlocalization exponent. Further theoretical work indicates that this d_ϕ should strictly be replaced by an 'internal length' dimension d_s relating the distance s_l between ramification points to the euclidean distance $l, s_l \propto l^{d_s}$ (Alexander 1989). The possible relation between d_s and the superlocalization of the wavefunction remains to be clarified. In terms of a 'ramification' dimension δ, connecting the modulus K_l to the internal length s_l, $K_l \propto s_l^{\delta-2}$, the expression for the Raman susceptibility can be rewritten (Alexander 1989)

$$I/n(\omega) \propto \omega^{-2+2(2-\bar{d})/(2-\delta)}. \tag{3}$$

From the measured slope, -0.37 ± 0.02, we find $d_s \approx 1.5$, meaning that the internal length is highly fractal. Also, the ramification dimension is $\delta \sim 1.15$, which is larger than 1, suggesting that the aerogels are indeed more ramified than percolation clusters. Also, δ is smaller than $\bar{d}$, as it should (Alexander 1989).

The density of states

It is presently of considerable interest to obtain a direct measurement of the DOS in a well-characterized fractal, in particular to determine the value of $\bar{d}$ without having to rely on mutual self-similarity and on scaling in the crossover region. One technique that can give a rather direct access to the DOS in the fractal régime is to observe incoherent neutron scattering from protons attached to silica particles (Freltoft *et al.* 1987). We have performed such a measurement on well-characterized neutrally reacted aerogels. The pure incoherent contribution was obtained by taking the difference of the signals derived from two identical samples, one with protons and the other with deuterons (Vacher *et al.* 1989 b). To this effect, we used the time-of-flight spectrometer MIBEMOL, at the Laboratoire Léon Brillouin. To achieve acceptable accumulation times, the resolution of the instrument was limited to 200 μeV in a first series of runs. This is only *ca.* 5 times smaller than the expected value of ω_{co2}, so that only the upper range of the fracton régime could be measured. However, these experiments allow us to determine the particle DOS. The results could be compared to a formula which has been derived for the DOS of small spheres (Baltes & Hilf 1973). This comparison requires no

† v indicates vertical polarization of the incident light. and H horizontal polarization of the scattered light.

adjustable parameters, except for the absolute value in the measurement, which becomes thus calibrated (Vacher *et al.* 1989 b). The DOS in the phonon régime can be calculated from the Brillouin results. The overall picture that emerges is presented in figure 8.

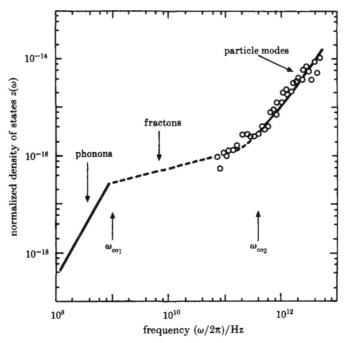

FIGURE 8. The DOS of the neutrally reacted gel N185. The solid line in the phonon regime is the Debye value calculated on the basis of the Brillouin scattering data. The points in the particle region are, up to a vertical displacement, obtained from neutron scattering measurements. The curve is from the theory of Baltes & Hilf (1973). The broken line is the conjectured fracton contribution.

In the particle régime, the effective slope of the DOS is around 1.5, because of the contribution of both surface (proportional to ω) and bulk (proportional to ω^2) modes. The curve is the fit to the theory of Baltes & Hilf. In the fracton régime, we have traced a line of slope $\bar{d} - 1 = 0.3$ through the measured points below ω_{co2}. It intercepts the phonon DOS near ω_{co1}. So far, the broken line should be considered as a conjecture. In particular we cannot decide yet whether or not there is a bump in the region of ω_{co1} (Aharony *et al.* 1985). However, as the proposed DOS is on an absolute scale, it can be used to calculate the specific heat C without any adjustable parameters. The behaviour that is found (Vacher *et al.* 1989 b) is in remarkable agreement with low-temperature measurements (Calemczuk *et al.* 1987). In particular, recent thermal results on neutrally reacted samples (Maynard *et al.* 1988) give a curve for C/T^3 that has exactly the same shape as our calculated curve (Vacher *et al.* 1989 b) and that agrees with it in absolute value within a factor of 2. This can be taken as strong evidence that fractons rather than two-level states control the low-temperature properties of these materials (Bon *et al.* 1987).

CONCLUSIONS

A first point worth stressing is that aerogels can be fractal over several orders of magnitude in length. Furthermore, as explained above, series of materials can be prepared that are mutually self-similar. This offers a way of investigating the dependence of properties on length scales by performing macroscopic measurements of these properties on series of samples, rather than performing microscopic measurements in function of length on one sample. The latter would be much more difficult. Suitably prepared aerogels appear to be the best available model system to investigate solid-state random fractals.

From the point of view of vibrations, a fully consistent picture has been obtained so far. It accounts for the Brillouin spectra near ω_{co1}, for the Raman spectra over the full range of existence of fractons, and for the particle modes observed in the inelastic neutron measurements of the DOS. Furthermore, the DOS that emerges is consistent with the low temperature thermal properties observed on the same materials. This accumulated evidence provides considerable support for fracton theories and for the scaling properties associated with these excitations.

The experiments revealed a number of dimensions: D, $\bar{d}$, d_s, and δ. Their values are summarized in table 2. For comparison, the expected dimensions for the infinite percolation cluster embedded in three-dimensional euclidean space are also listed in table 2. In the case of $\bar{d}$ two values are given, the first one corresponding to the so-called scalar or entropic elasticity (Alexander 1984), whereas the second is the tensorial result (Webman & Grest 1985). Related to these dimensions is also the exponent of the elastic modulus in function of the density, $K_l \propto \rho^{\tau/\beta}$. We use here the usual notation of percolation theory, with $K_l \propto (p-p_c)^\tau$, and $\rho \propto (p-p_c)^\beta$, where $p-p_c$ measures the distance above the percolation threshold. For the aerogels τ/β is related to D and $\bar{d}$, independently of the elasticity model (Courtens *et al.* 1987), by

$$\tau/\beta = \frac{\bar{d}+2D-D\bar{d}}{\bar{d}(3-D)}. \tag{4}$$

From the values of table 2, one finds $\tau/\beta \approx 3.8$, which also agrees with direct macroscopic measurements (Woignier *et al.* 1988). One notes that on percolation clusters, $\tau/\beta \approx 4$ in scalar elasticity, whereas $\tau/\beta \approx 9$ for bond-bending models (Deutscher *et al.* 1988). From this, and from the values in table 2, one would be

TABLE 2. VARIOUS DIMENSIONS

	neutrally reacted aerogels	infinite percolation clusters
D	2.4–2.46	2.5
$\bar{d}$	1.3	1.33, 0.9
d_s	1.5	—
δ	1.15	1

tempted to conclude that a neutrally reacted aerogel is nearly an infinite percolation cluster in which the large internal tensions induce a dominance of scalar elasticity. However, a more likely explanation is that these aerogels are in fact much more rigidly connected than infinite percolation clusters. The latter have a lot of very floppy endings. For the alcogel, such floppy arms are very likely to connect during ageing as well as during hypercritical drying, particularly because both processes are associated with shrinkage. A few additional connections would not affect appreciably the Hausdorff dimension, but it could modify considerably the elasticity. The increase of rigidity can very well raise $\bar{d}$ from 0.9 to 1.3, and also increase δ above 1.

The above review is based on joint work between the Laboratoire de Science des Matériaux Vitreux in Montpellier, the Laboratoire Léon Brillouin in Saclay, and the IBM Research Laboratory in Zürich. We gratefully acknowledge the contributions of many colleagues who participated to this research: J. Pelous, J. Phalippou, T. Woignier, G. Coddens and Y. Tsujimi. Many thanks are also expressed to Professor Alexander, Professor Orbach, Professor Aharony, Professor Entin-Wohlman and Professor Teixeira for many stimulating discussions.

REFERENCES

Aharony, A., Alexander, S., Entin-Wohlman, O. & Orbach, R. 1985 *Phys. Rev.* B 31, 2565.
Aharony, A., Alexander, S., Entin-Wohlman, O. & Orbach, R. 1987 *Phys. Rev. Lett.* 58, 132.
Alexander, S. & Orbach, R. 1982 *J. Phys. Lett.* 43, L625.
Alexander, S. 1984 *J. Phys.* 45, 1939.
Alexander, S. 1989 *Phys. Rev.* B (Submitted.)
Bale, H. D. & Schmidt, P. W. 1984 *Phys. Rev. Lett.* 53, 596.
Baltes, H. P. & Hilf, E. R. 1973 *Solid St. Commun.* 12, 369.
Bon, J., Bonjour, E., Calemczuk, R. & Salce, B. 1987 *J. Phys.* C 48, 483.
Calemczuk, R., de Goër, A. M., Salce, B., Maynard, R. & Zarembowitz, A. 1987 *Europhys. Lett.* 3, 1205.
Courtens, E., Pelous, J., Phalippou, J., Vacher, R. & Woignier, T. 1987 *Phys. Rev. Lett.* 58, 128.
Courtens, E., Vacher, R., Pelous, J. & Woignier, T. 1988 *Europhys. Lett.* 6, 245.
Deutscher, G., Maynard, R. & Parodi, O. 1988 *Europhys. Lett.* 6, 49.
Eden, M. 1961 In *Proceedings of the Fourth Berkeley Symposium on Mathematical Statistics and Probability* (ed. G. Neyman), vol. 4, pp. 223. Berkeley: University of California Press.
Freltoft, T., Kjems, J. & Richter, D. 1987 *Phys. Rev. Lett.* 59, 1212.
Fricke, J. 1985 *Aerogels*. Berlin: Springer-Verlag.
Guinier, A. 1956 *Théorie et technique de la radiocristallographie*. Paris: Dunod.
Iler, R. K. 1979 *The chemistry of silica*. New York: Wiley.
Kistler, S. S. 1932 *J. phys. Chem.* 36, 52.
Kolb, M., Botet, R. & Jullien, R. 1983 *Phys. Rev. Lett.* 51, 1123.
Maynard, R., Calemczuk, R., de Goër, A. M., Salce, B., Bon, J., Bonjour, E. & Bourret, A. 1989 In *Proceedings of the Second International Symposium on Aerogels* (ed. R. Vacher, J. Phalippou, J. Pelous & T. Woignier). *Revue Phys. appl.* 24, C4.
Meakin, P. 1983 a *Phys. Rev.* B 28, 5221.
Meakin, P. 1983 b *Phys. Rev. Lett.* 51, 1119.
Michard, F. & Zarembowitch, A. 1989 In *Proceedings of the Second International Symposium on Aerogels* (ed. R. Vacher, J. Phalippou, J. Pelous & T. Woignier) *Revue Phys. appl.* 24, C4.
Orbach, R. 1985 In *Scaling phenomena in disordered systems* (ed. R. Pynn & A. Skjeltorp), p. 335. New York: Plenum.

Polatsek, G. & Entin-Wohlman, O. 1988 *Phys. Rev* B 37, 7726

Schaefer, D. W. & Keefer, K. D. 1986 *Phys. Rev Lett.* **56**, 2199.

Schaefer, D. W., Hurd, A. J. & Glines, A. M 1988 In *Random fluctuations and pattern growth : experiments and models* (ed. H. E. Stanley & N. Ostrowski), p. 70. Dordrecht: Kluwer Academic Publishers.

Teixeira, J. 1986 *On growth and form* (ed. H E. Stanley & N Ostrowsky), p 145. Dordrecht. Nijhoff.

Tsujimi, Y , Courtens, E., Pelous, J. & Vacher, R. 1988 *Phys. Rev. Lett.* **60**, 2757.

Vacher, R., Woignier, T , Pelous, J. & Courtens, E. 1988a *Phys Rev.* B 37, 6500.

Vacher, R., Woignier, T., Phalippou, J., Pelous, J. & Courtens, E. 1988b *J non-cryst Solids.* **106**, 161.

Vacher, R., Phalippou, J., Pelous, J. & Woignier, T. (ed.) 1989a *Proceedings of the Second International Symposium on Aerogels, Montpellier, 21–23 September 1988. Revue Phys. appl.* **24 C4.**

Vacher, R , Woignier, T., Pelous, J., Coddens, G. & Courtens, E. 1989b *Europhys. Lett* **8**, 161.

Webman, I. & Grest, G. 1985 *Phys. Rev.* B 31, 1689.

Woignier, T., Phalippou, J., Sempere, R & Pelous, J 1988 *J. Phys.* **49**, 289.

Witten, T. & Sander, L. 1981 *Phys. Rev. Lett.* **47**, 1400.

Wong, P.-Z. & Bray, A. J. 1988 *Phys. Rev. Lett.* **60**, 1344.

Yakubo, K. & Nakayama, T. 1987 *Phys. Rev.* B 36, 8933.

Universality of fractal aggregates as probed by light scattering

By M. Y. Lin[1], H. M. Lindsay[2], D. A. Weitz[1], R. C. Ball[3], R. Klein[4] and P. Meakin[5]

[1] Exxon Research and Engineering, Rt. 22E, Annandale,
New Jersey 08801, U.S.A.
[2] Physics Department, Emory University, Atlanta, Georgia 30322, U.S.A.
[3] Cavendish Laboratory, Madingley Road, Cambridge CB3 0HE, U.K.
[4] Fakultät für Physik, Universität Konstanz, Konstanz, F.R.G.
[5] Central Research and Development Department,
Experimental Station, E. I. du Pont de Nemours and Company,
Wilmington, Delaware 19898, U.S.A.

Fractal colloid aggregates are studied with both static and dynamic light scattering. The dynamic light scattering data are scaled onto a single master curve, whose shape is sensitive to the structure of the aggregates and their mass distribution. By using the structure factor determined from computer-simulated aggregates, and including the effects of rotational diffusion, we predict the shape of the master curve for different cluster distributions. Excellent agreement is found between our predictions and the data for the two limiting régimes, diffusion-limited and reaction-limited colloid aggregation. Furthermore, using data from several completely different colloids, we find that the shapes of the master curves are identical for each régime. In addition, the cluster fractal dimensions and the aggregation kinetics are identical in each régime. This provides convincing experimental evidence of the universality of these two régimes of colloid aggregation.

Introduction

The aggregation of colloids has been the subject of scientific investigations for over 100 years. The past few years have seen considerable progress in our understanding of the complex physics that govern this process. A key to this recent success is the recognition that the structure of the colloidal aggregates exhibits scale invariance or dilation symmetry, and can be described as a fractal (Weitz & Oliveria 1984). This has afforded a quantitative description of the highly disordered structure of the clusters, which has in turn afforded a more detailed description of the kinetic growth process that forms these aggregates.

The class of colloid aggregation most widely considered is that which begins with a suspension of relatively monodisperse particles. Upon aggregation, these particles collide because of their Brownian motion and stick together to form larger clusters. The clusters themselves continue to diffuse, collide and form yet larger clusters. This process is called cluster–cluster aggregation, to distinguish it from other aggregation processes in which, for example, only single particles can diffuse, whereas the growing clusters remain stationary. Because all of the clusters participate in the growth, a complete characterization of the aggregation process must include a description of the distribution of the clusters formed. The kinetics of the aggregation can then be described in terms of the time evolution of this distribution. Both these kinetics and the shape of the cluster mass distribution are intrinsically related to the structure of the clusters that ultimately result.

Two distinct, limiting régimes of aggregation have been identified for colloids in solution (Weitz & Huang 1984; Weitz et al. 1985). These correspond directly to the classical régimes of rapid and slow aggregation that have been well established in the traditional colloid literature (Verwey & Overbeek 1948). The first of these régimes results in the most rapid aggregation possible, and occurs when the aggregation rate is limited solely by the time between the collisions of the clusters due to their diffusion. This régime is called diffusion-limited colloid aggregation (DLCA). The second limiting régime occurs when a large number of collisions are required before two particles can stick together, resulting in a much slower aggregation rate. This régime is called reaction-limited colloid aggregation (RLCA).

An elegant and quite detailed picture of cluster–cluster aggregation has evolved in recent years (Meakin 1988; Weitz et al. 1987). Computer simulations and analytical approaches have contributed to our theoretical understanding, and experimental studies have been conducted on several different colloidal systems. The description that emerges is a statistical one, both for the structure of the clusters in terms of their fractal dimensions, and for the cluster mass distribution and its evolution in time. As such, the description is independent of the details of the colloid system and should therefore be universal. Indeed, this is the great power of such a description: its applicability to a wide range of colloids. However, although this universality is appealing, it has never been verified experimentally. The purpose of this paper is to present an experimental technique that allows us to critically compare the aggregation behaviour of completely different colloids and to test the universality of the picture for aggregation that has evolved.

To compare critically the behaviour of completely different colloid systems, we use light scattering, both static and dynamic. Static light scattering is used to directly measure the fractal dimension, d_f, which characterizes the internal structure of the aggregates, as well as to determine the form of the cutoff function, which describes the finite extent of the clusters. Dynamic light scattering is used to measure the aggregation kinetics and the growth of the mean cluster size in time. In addition, we present a technique of scaling the results of dynamic light scattering onto a master curve. The shape of this master curve is very sensitive to many of the key features of the aggregation process: the shape of the cluster mass distribution; the fractal dimension of the aggregates; the anisotropy of the aggregates; and the cutoff function, which describes the bounds of the fractal structure. The master curves from entirely different colloids can be directly compared to evaluate critically the proposed universality of the aggregation processes.

To test the universality of colloid aggregation, we measure the behaviour of three completely different colloidal systems: gold, silica and polystyrene latex. Each colloid can be made to aggregate either very rapidly, under diffusion-limited kinetics, or very slowly, under reaction-limited kinetics. In this paper we demonstrate that the behaviour for each colloid is identical in each of the two limiting régimes. In each régime, the master curves for the colloids are indistinguishable. In each régime, the fractal dimensions of the aggregates are identical. In each régime, the aggregation kinetics for the colloids are the same. These results provide convincing experimental evidence proving that colloid aggregation is a universal process.

The remainder of this paper is organized as follows. In the next section we present a brief review of our current understanding of colloid aggregation and the experimental results that have been reported to date. In the following two sections, we develop the necessary description of the light scattering from fractal aggregates. We first discuss static light scattering, with particular emphasis on the cutoff function, which describes how the fractal structure of the clusters is bounded. We then discuss the interpretation of dynamic light scattering from fractal aggregates and show how the important contributions of rotational diffusion can be interpreted to provide a probe of the anisotropy of the aggregate structure. We also develop the master curves to compare dynamic light scattering results from different colloids. Finally, in the next section, we utilize static and dynamic light scattering and the master curves to compare the behaviour of the different colloids and verify that, in each régime, their aggregation is a universal process. A brief concluding section closes the paper.

REVIEW OF COLLOID AGGREGATION

The physics of colloid aggregation is ultimately determined by the nature of the interaction between approaching colloidal particles. Normally, the functional surface groups cause a repulsive interaction between two colloidal particles when they approach. If this repulsive energy barrier is much greater than $k_B T$, the colloid is stable against aggregation. Aggregation can be induced by reducing this barrier. If it is eliminated entirely, then only the very short-range, attractive Van der Waals interaction remains. This is the requirement for diffusion-limited colloid aggregation. By contrast, if the repulsive energy barrier is several $k_B T$, particles will still be able to stick to one another, but only after a large number of collisions. This is the requirement for reaction-limited colloid aggregation.

A complete characterization of each régime requires a description of the cluster structure, the aggregation kinetics and the shape of the cluster mass distribution (Weitz *et al.* 1985*b*, 1987). The cluster structure can be characterized by the fractal dimension, d_f, of the aggregates. The aggregation kinetics are described by the time evolution of the average cluster mass $\bar{M}$, which characterizes the distribution. These kinetics are characterized by an exponent, z, which determines the time evolution, $\bar{M} \sim t^z$. The cluster mass distribution typically exhibits dynamic scaling, in that its shape remains the same as it evolves in time (Vicsek & Family 1984). Thus, $N(M) = \Psi(M/\bar{M})/\bar{M}^2$, where $\Psi(M/\bar{M})$ is the scaling function that describes the shape of the cluster mass distribution. The behaviour of both régimes of aggregation has been studied theoretically and by means of computer simulations. In addition, there have been experimental studies with several different types of colloids of each régime.

For DLCA, computer simulations (Meakin 1983; Kolb *et al.* 1983) have shown that the clusters formed are fractal, with $d_f \approx 1.8$. The fractal dimension of DLCA aggregates has been measured for several different colloids, including gold (Weitz & Oliveria 1984), silica (Aubert & Cannell 1986) and polystyrene (Matsushita *et al.* 1986). Several different experimental techniques have been used, including the analysis of transmission electron micrographs (TEM), light scattering and X-ray scattering (Dimon *et al.* 1986). In all cases a value of $d_f \approx 1.85 \pm 0.1$ is obtained, consistent with the computer simulations.

The time dependence of the characteristic cluster mass can be determined analytically by means of the Smoluchowski equations (Van Dongen & Ernst 1985). For DLCA, a linear dependence is predicted, so that $z = 1$ and $\bar{M} = (t/t_0)^z$. Here the characteristic time is given by $t_0 = 3\eta/(8k_B T N_0)$, with η the viscosity of the fluid and N_0 the initial particle concentration. Computer simulations also find $z = 1$ (Meakin *et al.* 1985). The kinetics of DLCA have been measured for colloidal gold and are found to be consistent with the theoretical predictions of $z = 1$. (Weitz *et al.* 1984).

The cluster mass distribution, $N(M)$, can be determined analytically with the Smoluchowski rate equations (Cohen & Benedek 1982). For DLCA, it is given to a good approximation by the exponential form

$$N(M) = \frac{N_T}{\bar{M}} \left[1 - \frac{1}{\bar{M}} \right]^{M-1}, \tag{1}$$

where $N_T = \sum N(M)$, the total number of clusters. The average cluster mass is defined explicitly as $\bar{M} = \sum MN(M)/N_T$. Computer simulations of DLCA predict cluster mass distributions that are in good agreement with this result (Meakin *et al.* 1985). Experimental measurements with TEM counting methods with colloidal gold also support this result, although the statistical accuracy of this type of counting technique is inherently limited (Weitz & Lin 1986).

The behaviour of RLCA is strikingly different. Computer simulations again confirm that the clusters are fractal, with $d_f \approx 2.1$ (Brown & Ball 1985). Experimentally, clusters formed by RLCA are also found to have $d_f \approx 2.1 \pm 0.1$, for measurements made on gold, silica (Schaeffer *et al.* 1984) and polystyrene colloids.

The aggregation kinetics for RLCA are also very different than for DLCA. They can be determined analytically from the Smoluchowski equations and are best described as exponential, $\bar{M} \sim e^{At}$, where A is a constant dependent on the sticking probability and the time between collisions. Dynamic light scattering measurements on colloidal gold and static light scattering measurements on colloidal silica also find an exponential growth of the characteristic cluster mass.

The cluster mass distribution for RLCA is also very different from that of DLCA. A solution of the Smoluchowski equations (Ball *et al.* 1987) suggests that it is best described as a power-law up to a cutoff mass, after which it again decreases exponentially, $N(M) \sim M^{-\tau} e^{-M/M_c}$. There is some disagreement about the value of τ, with the approach using the Smoluchowski equations suggesting that $\tau = 1.5$, whereas computer simulations (Meakin & Family 1987) find a somewhat larger value of $\tau \approx 1.75$. All experimental measurements reported to date are consistent with the power-law shape of the cluster mass distribution, but there is again some apparent disagreement in the value of τ obtained. Early measurements of antigen–antibody aggregation of polystyrene spheres found $\tau = 1.5$, although the measurements were limited to cluster masses of $M \approx 10$ (Von Schultess *et al.* 1980). Measurements on colloidal gold with TEM counting also found $\tau \approx 1.5 \pm 0.1$, within the inherent statistical limitations of this technique. By contrast, an interpretation of dynamic light scattering data from colloidal silica suggested a higher value of $\tau \approx 1.9$ (Martin & Leyvraz 1986).

STATIC LIGHT SCATTERING

Knowledge of the intensity of the light scattered from a single colloidal aggregate is crucial for the interpretation of all light-scattering data. In particular, we must know its dependence on both scattering vector and cluster mass. We express the scattering from a single cluster as $M^2 S(qR_g)$, where the structure factor behaves as

$$S(qR_g) \simeq \begin{cases} 1 & \text{for } qR_g \ll 1 \\ (qR_g)^{-d_t} & \text{for } qR_g \gg 1. \end{cases} \tag{2}$$

Here the scattering vector is $q = (4\pi n/\lambda)\sin(\frac{1}{2}\theta)$, with n the index of refraction of the fluid, λ the wavelength and θ the scattering angle. We have implicitly assumed that there is only one characteristic length for each cluster, given by R_g, so that the structure factor is a function of qR_g. The radius of gyration of each cluster can be related to its mass through the fractal scaling of the clusters

$$M = (R_g/a)^{d_t}, \tag{3}$$

where a is the radius of a single particle.

At small qR_g, the internal structure of the cluster is not resolved and it scatters coherently, so that the intensity scales as M^2. By contrast, at larger qR_g, the fractal structure is resolved, resulting in the familiar q^{-d_t} dependence of the scattered intensity. However, the cluster is now sufficiently large that the scattering intensity no longer adds completely coherently, and the intensity now scales as M. This latter régime allows the measurement of d_f directly.

Although equation (2) provides the asymptotic limits, the full structure factor must be determined to compare with experimental data. This requires knowledge of the crossover behaviour of the structure factor at $qR_g \approx 1$, which is determined by the way the cluster structure is bounded at the edges of the clusters. The boundaries of fractal clusters are rather sharp, and computer simulations suggest that the density correlation function for cluster–cluster aggregates can be characterized by a stretched exponential cutoff (Mountain & Mulholland 1988). To determine this crossover behaviour of $S(qR_g)$, we use computer-simulated clusters and calculate their structure factors directly

$$M^2 S(qR_g) = \left\langle \left| \sum_j e^{iq \cdot b_j} \right|^2 \right\rangle, \tag{4}$$

where the summation extends over all M particles in the cluster, b_j is a vector signifying the position of the jth particle and $\langle \ \rangle$ signifies orientational averaging, which is achieved by performing the calculation repeatedly for many different orientations of the cluster. Typical results obtained from a large number of computer-simulated RLCA clusters are shown by the crosses in figure 1 a, where we plot $S(qR_g)$ as a function of qR_g.

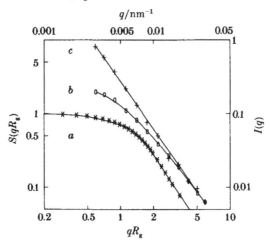

FIGURE 1. Static light scattering intensity from fractal aggregates formed by reaction-limited aggregation. Lines: a, average structure factor calculated from computer-simulated clusters and plotted against qR_g (lower scale); b and c, experimental data obtained from gold RLCA aggregates at different times in the aggregation process. The solid lines are the fits using the structure factor in (a) and assuming a power-law cluster mass distribution with $\tau = 1.5$.

To describe the structure factor of these clusters, it is convenient to use the functional form

$$S(qR_g) = \left(1 + \frac{8}{3d_f}(qR_g)^2 + C_2(qR_g)^4 + C_3(qR_g)^6 + C_4(qR_g)^8\right)^{-\frac{1}{8}d_f}, \quad (5)$$

where C_2, C_3 and C_4 are fitting parameters. This functional form has the correct asymptotic limits for $qR_g \ll 1$ and $qR_g \gg 1$. In addition, it has the correct initial behaviour for small qR_g, given by the Guinier form, $S(qR_g) \approx 1 - \frac{1}{3}(qR_g)^2$. The higher-order terms in qR_g are required to account for the crossover behaviour of the structure factor around $qR_g \approx 1$. We emphasize, however, that equation (6) is simply a parameterization of the calculated structure factor, and there is no physics in its choice. This form describes the calculated structure factor very well as shown by the solid line in figure 1 a, where we have used $C_2 = 3.13$, $C_3 = -2.58$ and $C_4 = 0.95$. We use $d_f = 2.08$, as measured by the dependence of the mass on radius of gyration for the computer-simulated clusters.

To compare to experimental data, we must sum the scattering intensity of the individual clusters over the distribution

$$I(q) = \sum N(M) M^2 S(qR_g). \quad (6)$$

The structure factor that describes the computer-generated clusters also describes the experimental data for RLCA very well. This is shown by figure 1 (b, c), which are data obtained from colloidal gold at different times during a reaction-limited aggregation process. The data were fit to equation (5) by using a power-law cluster mass distribution with $\tau = 1.5$. The only fitting parameter is the cutoff mass of the exponential in the distribution function, M_c. As can be seen by the solid lines, the agreement between the experimental data and the calculation is excellent. By contrast, using other suggested forms for the structure factor (Sinha *et al.* 1984; Wiltzius 1987), gives markedly poorer agreement with the data.

It should be noted from figure 1 that the crossover of the experimental data is considerably less sharp than that of the structure factor of the individual clusters. This is because of the effects of the cluster mass distribution, which causes the apparent rounding in the total scattered intensity. Thus, the total scattered intensity can be described by, for example, a Fisher–Burford form (Wiltzius 1987), but the characteristic cluster radius determined by such a fit has no direct relation with M_c.

Static light scattering can be used to determine the fractal dimension of the clusters directly, provided they are large enough that most of them have $qR_g > 1$ for the wavevectors probed. By contrast, if the clusters are small enough that the crossover can be resolved, $\bar{M}$ can be determined from the static scattering. As seen by figure 1, with a power-law cluster mass distribution the asymptotic q^{-d_f} behaviour of the fractal region is only achieved at values of q much larger than the crossover range. However, we cannot independently obtain information about the details of the shape of the cluster mass distribution from the static scattering intensity. For example, in figure 1 (b, c), we can obtain equally good fit to the data by assuming a different value for τ.

DYNAMIC LIGHT SCATTERING

Dynamic light scattering provides an additional, very useful experimental probe of fractal colloid aggregates. It measures the temporal fluctuations in the scattered light intensity resulting from the diffusive motion of the clusters. From the decay of the autocorrelation function of these fluctuations, it is possible to determine the size of the characteristic cluster in the distribution (Berne & Pecora 1976). However, to do so, we must consider not only the translational diffusion of the clusters, but also their rotational diffusion when the size of the clusters is larger than q^{-1}. Because the internal fractal structure of the clusters is resolved, the scattered intensity is strongly dependent on the orientation of the cluster, and consequently fluctuates as the aggregate undergoes rotational diffusion. A calculation of the increase in the decay rate of the intensity autocorrelation function involves a determination of the anisotropy of the clusters. Then the q-dependence of the measured autocorrelation function can be used to probe this anisotropy (Lindsay *et al.* 1987, 1988 a).

We can account for the increase in the decay rate due to the contribution of rotational diffusion by means of an effective diffusion coefficient, $\Gamma = q^2 D_{\mathrm{eff}}$. For a single cluster, we can obtain an approximation for the functional form for D_{eff} in terms of the static structure factor (R. C. Ball, personal communication)

$$\frac{D_{\mathrm{eff}}}{D} = 1 + \frac{1}{2\beta^2}\left[1 + \frac{3\partial \ln S(qR_g)}{\partial(qR_g)^2}\right], \tag{7}$$

where D is the Stokes–Einstein translational diffusion coefficient and $\beta = R_H/R_g$, the ratio between the hydrodynamic radius and the radius of gyration. We make the approximation that a single hydrodynamic radius can be used for both translational and rotational diffusion. Insight into the behaviour of D_{eff} can be obtained by examining the asymptotic forms of equation (7). When $qR_g < 1$, we can use the Guinier expansion of $S(qR_g) = 1 - \frac{1}{3}(qR_g)^2$, to obtain $D_{eff}/D = 1$, so that rotational diffusion makes no contribution, as expected. For $qR_g \gg 1$, the derivative term decreases as $(qR_g)^{-1}$, so that $D_{eff}/D = 1 + 1/(2\beta)^2$, and is a constant. Thus D_{eff} varies only when $qR_g \approx 1$, and is a constant in the two asymptotic limits. It is also possible to determine D_{eff} from the more exact calculations of the structural anisotropy, and good agreement is obtained with the predictions of equation (7) (Lindsay et al. 1988 b). A critical feature obtained with both methods is that D_{eff} is a function of qR_g. This means that for dynamic light scattering, as for static light scattering, the clusters can be characterized by a single length scale, R_g.

To compare with experimental data, we must average D_{eff} over the cluster distribution. Here, we calculate the measured first cumulant of the field autocorrelation function, or the logarithmic derivative at zero time, $\bar{\Gamma}_1$, which, when divided by q^2, gives an average effective diffusion coefficient, $\bar{D}_{eff}$ (Berne & Pecora 1976). It can be expressed as

$$\bar{D}_{eff} = \frac{\bar{\Gamma}_1}{q^2} = \frac{\sum N(M) M^2 S(qR_g) D_{eff}}{\sum N(M) M^2 S(qR_g)}. \tag{8}$$

Here the effective diffusion coefficients are weighted by the scattering intensity and the number of clusters of each mass. The fractal scaling of the clusters, equation (3), can be used to relate R_g to the mass of the clusters to carry out the summations. We note that the orientationally averaged static structure factor is used as the intensity weighting, even though rotational diffusion depends on the lack of orientational averaging in determining the fluctuations of the scattered intensity. The effects of rotational diffusion are included in equation (8) only through D_{eff}. This intensity weighting is implicit in equation (7) for D_{eff} and is one of the advantages of this approximation.

Even for a relatively monodisperse cluster mass distribution, such as that obtained for DLCA, the effects of rotational diffusion cause the measured D_{eff} to be dependent on q. This additional dependence on q is observed experimentally and is well described using the form of D_{eff} in equation (8) (Lindsay et al. 1988 a, b).

Finally, we note that equation (8) can be used to obtain an experimental measure of β, the ratio between the hydrodynamic radius, which determines the diffusion coefficient of the clusters, and the radius of gyration, which determines their size for static light scattering. This is done by measuring the static light scattering intensity and fitting to equation (5) to obtain $\bar{M}$ for the cluster mass distribution. This same $\bar{M}$ can then be used in equation (8) to calculate $\bar{\Gamma}_1$. This can be done at sufficiently small q that rotational diffusion does not contribute, so that the only unknown parameter is β. We find $\beta = 0.93$ for DLCA and $\beta = 1.0$ for RLCA (Lin et al. 1988). These values are in agreement with calculations (Chen et al. 1984; Hess et al. 1986) and other measurements (Wiltzius 1987; Pusey et al. 1987).

MASTER CURVES FOR DYNAMIC LIGHT SCATTERING

There is considerably more information obtained in the measurements of $\bar{D}_{\text{eff}}$ if the q dependence is also determined. This can be seen by examining equation (8), which can be viewed as a measure of a moment of the cluster mass distribution. Because the weighting of the structure factor of the clusters changes with qR_g, the moment will also depend on q. If a sufficiently small q can be achieved that all clusters have $qR_g < 1$, then the intensity weighting for each cluster is M^2. Because rotational diffusion will not contribute, the diffusion coefficient adds a factor proportional to M^{-1/d_f}, so the moment of the cluster mass distribution function would be $2 - 1/d_f$. However, as q increases, some of the clusters become larger than q^{-1}, and the moment measured is no longer so straightforward. At very large q, the result reverts to a simple moment, where now the intensity from each cluster contributes only a factor of M in the weighting. In addition, although rotational diffusion contributes to D_{eff}, the scaling with cluster mass is changed. Thus the moment measured is $1 - 1/d_f$. However, to reach this measure would require $qa > 1$, which is not achieved for any of the colloids studied here.

In our experiments, the range of q that is accessible is limited. However, we can take advantage of the dynamic scaling of the cluster mass distribution to greatly extend the range over which we can measure the q dependence of $\bar{D}_{\text{eff}}$. This is illustrated in figure $2a$, where we plot the $\bar{D}_{\text{eff}}$ calculated with equation (8) for RLCA clusters using $M_c = 10^4$, $a = 7.5$ nm and $\tau = 1.5$. The experimentally accessible range of q extends over about one decade of the three shown. The effects of further aggregation are shown in figure $2b$, c for $M_c = 10^5$ and 10^6. Because the cluster mass distribution exhibits dynamic scaling and its shape remains the same as the aggregation proceeds, the shape of the curves for $\bar{D}_{\text{eff}}$ also remains the same.

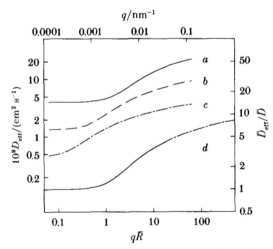

FIGURE 2 Calculated values of $\bar{D}_{\text{eff}}$ as a function of q for (a) $M_c = 10^4$, (b) $M_c = 10^5$ and (c) $M_c = 10^6$. For these curves the q scale is on the top of the plot and the $\bar{D}_{\text{eff}}$ scale is to the left. The experimentally accessible range of q is from 0.002 to 0.03 nm^{-1}. The data are scaled onto a master curve (d) whose scales are on the bottom and the right of the figure

Each curve is characterized by a single mass, M_c, and can be represented by a single average diffusion coefficient, $\bar{D}$, which would be measured at $q = 0$, and the corresponding hydrodynamic radius, $\bar{R}$. Thus when we plot $\bar{D}_{\text{eff}}/\bar{D}$ as a function of $q\bar{R}$, all the data lie on a single master curve, as shown in figure $2d$.

The scaling of the curves of $\bar{D}_{\text{eff}}$ allows us to greatly extend the range of our measurements despite the experimental limitations in available q. This is accomplished by repeating the measurement at different times, which correspond to different $\bar{M}$ and hence different parts of the master curve. Then by scaling the data, the shape of the complete master curve can be determined. In practice, the lowest values of q accessible are not always sufficient to measure $\bar{D}$ directly. Instead, the data must be scaled experimentally. Because $\bar{D}^{-1} \propto \bar{R}$, this scaling entails shifting each data set along the diagonal of a logarithmic plot of $\bar{D}_{\text{eff}}$ as a function of q until it overlaps with the data set obtained at another time. We note that this technique utilizes the q dependence of data obtained at different times and can thereby provide new information.

The shape of this master curve provides us with a great deal of information about the aggregation process. Its shape depends critically on the cluster mass distribution. Furthermore, its shape will be sensitive to the contribution of rotational diffusion to D_{eff}, and thus will reflect the anisotropy of the clusters. As the behaviour of the master curve results explicitly from the change in the structure factor of the clusters, its shape will directly reflect the shape of the structure factor and, in particular, the crossover region where $qR_g \approx 1$. Finally, the scaling factors required to shift the data onto the master curve provide a measure of $\bar{M}$ or $\bar{R}$, and their time dependence will directly reflect the aggregation kinetics. However, all features specific to the individual colloid, such as the single particle radius, have been scaled out of the master curve. Thus these master curves provide an extremely sensitive means of comparing the aggregation behaviour of completely different colloids.

UNIVERSALITY OF COLLOID AGGREGATION

We can use the light scattering results of the previous sections to critically compare the behaviour of different colloids to investigate the universality of the two limiting régimes of colloid aggregation. We do this by using static light scattering to determine the fractal dimensions of the clusters and by using dynamic light scattering measurements to obtain a master curve for each colloid, as well as a measure of the aggregation kinetics through the time dependence of the characteristic cluster mass. The results of different colloids can then be compared in each régime.

We use three completely different colloids: gold, silica and polystyrene latex. Each colloid is composed of a different material; each colloid is initially stabilized by completely different functional groups on its surface; the aggregation for each colloid is initiated in a different manner; the interparticle bonds in the aggregates for each colloid are different; and each colloid has a different primary particle size. However, each colloid can be made to aggregate by either diffusion-limited or reaction-limited kinetics.

The colloidal gold has a particle radius of $a = 7.5$ nm and an initial volume fraction of $\phi_0 = 10^{-6}$. It is stabilized by citrate ions adsorbed on the surface. The aggregation is initiated by addition of pyridine, which displaces the charged ions, reducing the repulsive barrier between the particles. The amount of pyridine added determines the aggregation rate: for DLCA, the pyridine concentration is 10^{-2} M, whereas for RLCA, it is about 10^{-5} M. The interparticle bonds are metallic.

The colloidal silica used is Ludox SM obtained from DuPont. It has particles with $a = 3.5$ nm, and is diluted to $\phi_0 = 10^{-6}$. It is initially stabilized by OH$^-$ or SiO$^-$ on the surface. The pH is kept at 11 or more by addition of NaOH and the aggregation is initiated by addition of NaCl, which reduces the Debye–Hückel screening length, thereby reducing the repulsive barrier between the particles. For DLCA, the salt concentration is 0.9 M, whereas for RLCA, it is 0.6 M. The interparticle bonds are believed to be siloxane bonds.

The polystyrene latex has $a = 19$ nm and is diluted to $\phi_0 = 10^{-6}$. It is initially stabilized by charged carboxylic acid groups on the surface of the particles. Addition of HCl to a concentration of 1.2 M is used to neutralize the surface charges and decrease the screening length to initiate the aggregation for DLCA. For RLCA, NaCl is added to a concentration of 0.2 M, to reduce the screening length and initiate the aggregation. The particle surfaces deform on bonding, leading to large Van der Waals interactions between the bound particles.

We begin by investigating the régime of diffusion-limited colloid aggregation. In figure 3, we show a logarithmic plot of the static light scattering obtained from the three colloids aggregated by DLCA, when the cluster sizes are sufficiently large that the scattering is in the fractal region for all the q accessible experimentally. The solid lines are fits to the data and give the fractal dimensions of the aggregates: for gold, $d_f = 1.86$; for silica, $d_f = 1.85$; and for polystyrene, $d_f = 1.86$. These values are indistinguishable to within experimental error.

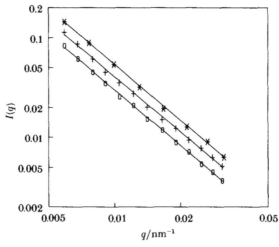

FIGURE 3. Static light scattering from DLCA aggregates of gold (o), silica (+) and polystyrene (*). The measured fractal dimensions are gold, $d_f = 1.86$; silica, $d_f = 1.85$; and polystyrene, $d_f = 1.86$.

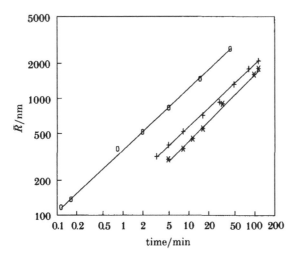

FIGURE 4. Kinetics of DLCA aggregation for gold (○), silica (+) and polystyrene (∗). From the slopes the dynamic exponents are: gold, $z = 0.99$; silica, $z = 1.00$; and polystyrene, $z = 1.05$.

To obtain $\bar{D}_{eff}$ the first cumulant of the autocorrelation function was measured repeatedly as a function of q as the aggregation proceeded. At each q, $\bar{D}_{eff}$ was found to decrease with time as a power-law, and this trend was used to interpolate between the data to obtain sets of values measured at the same time, as required for the scaling procedure. The time dependence of $\bar{R}$ determined from the scaling factors is shown in a logarithmic plot in figure 4. The data all exhibit the power-law behaviour expected for DLCA. The solid lines are fits to the data and their slopes, α are related to the dynamic exponent, $R = a\bar{M}^{1/d_f} = a(t/t_0)^{z/d_f}$. Using the fractal dimensions measured with static light scattering, we obtain $z = 0.99$ for gold, $z = 1.00$ for silica and $z = 1.05$ for polystyrene. Thus, within experimental error, $z = 1.0$ for all the colloids. In addition, the characteristic times measured from the data are very close to those predicted for each of the colloids. These results confirm that the kinetics for DLCA are identical for all the colloids.

We plot the DLCA master curves for all three colloids in figure 5. The three sets of data are indistinguishable, confirming that the aggregation behaviour of the colloids is identical. We emphasize that this result is independent of any interpretation of the data. The solid line through the data is the calculated master curve from equation (8) and the cluster mass distribution given by equation (1) from the Smoluchowski equations applied to DLCA. It is in good agreement with the data. The q-dependence of the master curve is almost entirely a result of the contribution of rotational diffusion, which affects the measured $\bar{D}_{eff}/\bar{D}$ only around $q\bar{R} \approx 1$. However, this is the régime that is probed by most of the experimental measurements.

We can make exactly the same comparisons for the colloids aggregated by RLCA. The static light scattering from the three colloids in the fractal régime is shown in a logarithmic plot in figure 6. The solid lines are fits to the data, and give the

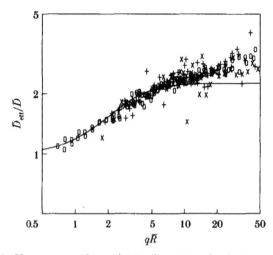

FIGURE 5. Master curves for gold (○), silica (+) and polystyrene (×) for DLCA, compared with the predicted shape shown by the solid line

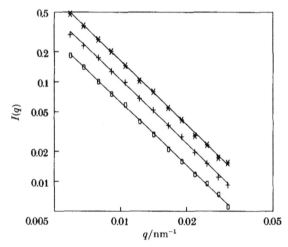

FIGURE 6. Static light scattering from RLCA aggregates of gold (○), silica (+) and polystyrene (∗). The measured fractal dimensions are. gold, $d_f = 2.10$; silica, $d_f = 2\ 12$; and polystyrene, $d_f = 2\ 13$

fractal dimension of the aggregates: for gold $d_f = 2.10$; for silica, $d_f = 2.12$; and for polystyrene $d_f = 2.13$. Again the results are identical to within experimental error.

The kinetics of the aggregation processes are shown in figure 7, where we plot the time dependence of $\bar{R}$ obtained from the scaling of the data sets onto the master curve. The kinetics of each colloid are exponential, as shown in the semi-logarithmic plot in figure 7.

The master curves for the three colloids are plotted in figure 8. The data sets are

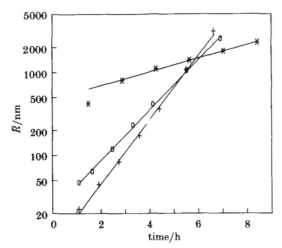

FIGURE 7. Kinetics of RLCA aggregation for gold (o), silica (+) and polystyrene (*).
All the colloids display exponential kinetics for RLCA.

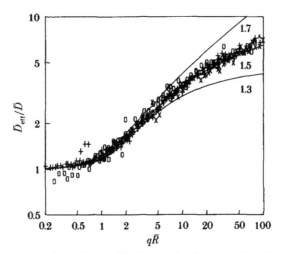

FIGURE 8. Master curves for gold (o), silica (+) and polystyrene (×) for RLCA. The solid line
show the shapes predicted with different cluster mass exponents. The value of $\tau = 1.5$
provides the best agreement with the data for all the colloids.

again indistinguishable, confirming that the aggregation behaviour in RLCA is
identical for all the colloids. There is considerably more variation of $\bar{D}_{\mathrm{eff}}/\bar{D}$ with
$q\bar{R}$ than there is for DLCA. This results from the power-law shape of the cluster
mass distribution, which leads to an additional q-dependence in $\bar{D}_{\mathrm{eff}}$. However, the
master curve data provide an independent measure of the shape of the cluster
mass distribution. The solid lines show the predictions of equation (8) assuming a
power-law cluster mass distribution, for three different values of τ. The data are
best described by the curve calculated for $\tau = 1.5$.

In obtaining the master curves for the RLCA data, there are several precautions that must be taken. To apply the concepts of aggregation developed above, the initial concentrations must be low enough that the volume fraction of clusters is well below the gel point for all times measured, which requires $\bar{R}/a \ll \phi_0^{1/(d_f-3)}$. For our data, where $\bar{R}/a$ can exceed 10^3, using $\phi_0 \approx 10^{-6}$ ensures that gelation is not approached. If gelation is approached, the cluster mass exponent must increase to $\tau \geq 2$, and the shape of the master curve obtained will change. Previous data obtained from silica were presumably obtained by using higher initial concentrations, and the gel point may have been approached (Martin 1987), giving rise to the different shape of the master curve that was observed. In addition, for RLCA, the aggregation time is so long, and the cluster mass distribution so broad, that precautions must be taken to avoid differential settling due to gravity. In our experiments, this was achieved by using a long sample container filled to the top, and inverting it every 15 min so that on average there was no gravitational settling. Failure to do so results in marked changes in the shape of the master curve obtained.

The data presented here provides striking evidence in support of the universality of each of the two régimes of colloid aggregation. The static scattering demonstrates that the clusters formed have identical fractal dimensions in each régime, with $d_f = 1.85 \pm 0.05$ for DLCA and $d_f = 2.11 \pm 0.05$ for RLCA. The time dependence of $\bar{R}$ demonstrates that the kinetics behave in identical fashions in each régime, with $\bar{M}$ growing linearly in time for DLCA and exponentially in time for RLCA. However, the most convincing evidence of the universality of the colloid aggregation is the similarities of the master curves for all the colloids in each régime. The shape of the master curve is very sensitively dependent on the features of the colloid aggregation: the cluster mass distribution, the structure factor and the anisotropy of the clusters through the contribution of rotational diffusion. The fact that the shapes of the master curves are identical for each colloid implies that each of these features is the same for each colloid. This convincingly demonstrates the universality of colloid aggregation.

CONCLUSIONS

In this paper, the process of colloid aggregation is studied by using both static and dynamic light scattering. The focus is on verifying the universality of our current picture of colloid aggregation. We investigate the two limiting régimes, diffusion-limited and reaction-limited colloid aggregation. For each régime, we compare the behaviour of three completely different colloids: gold, silica and polystyrene. Static light scattering is used to measure the fractal dimension of the aggregates. Dynamic light scattering is used to measure the kinetics of aggregation. The dynamic light scattering data are also scaled onto a single master curve whose shape is shown to be a very sensitive measure of the details of the aggregation process, including the structure of the clusters and their cluster mass distribution. The master curves for the different colloids can be compared to critically test for universal behaviour.

For DLCA, we find that all the colloids have $d_f = 1.85 \pm 0.05$. For all the colloids, the kinetics of aggregation are described by a power-law behaviour in time, with the average cluster radius growing as $\bar{R} \sim (t/t_0)^{z/d_f}$, where t_0 is a characteristic time that depends on the initial colloid concentration and the particle radius. Using the static light scattering measurement of d_f, we find that all the colloids have $z = 1.0 \pm 0.05$. The master curves for all the colloids are indistinguishable, and their shape is in excellent agreement with our prediction using the cluster mass distribution that is expected for DLCA. This cluster mass distribution is constant in mass at low masses and decreases exponentially above the average mass, $\bar{M}$.

For RLCA, we find that all the colloids have $d_f = 2.11 \pm 0.05$. For all the colloids, the kinetics of aggregation is exponential in time. The time constant is a function of the sticking probability of two particles and the time between collisions. The master curves for all three colloids are again indistinguishable. Their shape is in excellent agreement with the prediction using a power-law cluster mass distribution, with an exponent of $\tau = 1.5 \pm 0.05$.

These results confirm that each of the two limiting régimes of colloid aggregation exhibit distinct, yet universal, behaviour. The most conclusive evidence is the similarity in shape of the master curves for the three colloids in each régime. This similarity implies that the cluster structure, the cluster anisotropy and the cluster mass distribution are all universal for each régime.

REFERENCES

Aubert, C. & Cannell, D. S. 1986 Restructuring of colloidal silica aggregates. *Phys. Rev. Lett.* **56**, 738.

Ball, R. C., Weitz, D. A., Witten, T. A. & Leyvraz, F. 1987 Universal kinetics in reaction-limited aggregation. *Phys. Rev. Lett.* **58**, 274.

Berne, B. J. & Pecora, R. 1976 *Dynamic light scattering.* New York: Wiley-Interscience.

Brown, W. D. & Ball, R. C. 1985 Computer simulation of chemically limited aggregation. *J. Phys.* A **18**, L517.

Chen, Z.-Y., Deutsch, J. M. & Meakin, P. 1984 Translational friction coefficient of diffusion limited aggregates. *J. chem. Phys.* **80**, 2982.

Cohen, R. J. & Benedek, G. B. 1982 Equilibrium and kinetic theory of polymerization and the sol-gel transition. *J. chem. Phys.* **86**, 3696.

Dimon, P., Sinha, S. K., Weitz, D. A., Safinya, C. R., Smith, G. S., Varady, W. A. & Lindsay, H. M. 1986 Structure of aggregated colloids. *Phys. Rev. Lett.* **57**, 595.

Hess, W., Frisch, H. L. & Klein, R. 1986 On the hydrodynamic behavior of colloidal aggregates. *Z. Phys.* B **64**, 65.

Kolb, M., Botet, R. & Jullien, R. 1983 Scaling of kinetically growing clusters. *Phys. Rev. Lett.* **51**, 1123.

Lin, M. Y., Lindsay, H. M., Weitz, D. A., Klein, R., Ball, R. C. & Meakin, P. 1989 (Preprint.)

Lindsay, H. M., Lin, M. Y., Weitz, D. A., Sheng, P., Chen, Z., Klein, R. & Meakin, P. 1987 Properties of fractal colloid aggregates. *Faraday Discuss. chem. Soc.* **83**, 153.

Lindsay, H. M., Klein, R., Weitz, D. A., Lin, M. Y. & Meakin, P. 1988a The effect of rotational diffusion on quasielastic light scattering from fractal colloidal aggregates. *Phys. Rev.* A **38**, 2614–2626.

Lindsay, H. M., Lin, M. Y., Weitz, D. A., Ball, R. C., Klein, R. & Meakin, P. 1988b Light scattering from fractal colloid aggregates. In *Proc. of Topical Meeting on Photon Correlation Techniques and Applications* (ed. A. Smart & J. Abbiss). Washington, D.C.: Optical Society of America.

Martin, J. E. & Leyvraz, F. 1986 Quasielastic-scattering linewidths and relaxation times for surface and mass fractals. *Phys. Rev.* A **34**, 2346.

Martin, J. E. 1987 Slow aggregation of colloidal silica. *Phys. Rev.* A **36**, 3415.

Matsushita, M , Hayakawa, Y , Sumida, K & Sawada, Y 1986 Experimental investigation of the fractality and kinetics of aggregation In *Science on Form · Proceedings of the First International Conference for Science on Form* (ed Y Kato, R Takaki & J Toriwaki). Tokyo: KTK Scientific Publishers.

Meakin, P. 1983 Formation of fractal clusters and networks by irreversible diffusion-limited aggregation. *Phys. Rev. Lett* **51**, 1119.

Meakin, P , Vicsek, T. & Family, F 1985 Dynamic cluster size distribution in cluster-cluster aggregation. *Phys. Rev.* B **31**, 564

Meakin, P. & Family, F 1987 Structure and dynamics of reaction limited aggregation. *Phys. Rev.* A **36**, 5498

Meakin, P 1988 Fractal aggregates and their fractal measures In *Phase translations*, vol. 12 (ed. J L. Liebowitz). New York: Academic Press.

Mountain, R D. & Mulholland, G. W. 1988 Light scattering from simulated smoke agglomerates *Langmuir* **4**, 1321

Pusey, P N , Rarity, J. G , Klein, R & Weitz, D. A. 1987 Comment on 'hydrodynamic behavior of fractal aggregates', by P Wiltzius. *Phys. Rev. Lett.* **59**, 2122.

Schaeffer, D. W., Martin, J. E., Wiltzius, P. & Cannell, D. S 1984 Fractal geometry of colloidal aggregates. *Phys. Rev. Lett.* **52**, 2371.

Sinha, S. K , Frelthoft, T. & Kjems, J. 1984 Observation of power-law correlations in silica-particle aggregates by small angle neutron scattering. In *Kinetics of aggregation and gelation* (ed. F Family & D. P. Landau). Amsterdam · North-Holland.

Van Dongen, G J. & Ernst, M. H 1985 Dynamic scaling in the kinetics of clustering *Phys. Rev. Lett* **54**, 1396

Verwey, E. J. W. & Overbeek, J T. G. 1948 *Theory of the stability of lyophobic colloids.* Amsterdam. Elsevier.

Vicsek, T & Family, F 1984 Dynamic scaling for aggregation of clusters. *Phys. Rev. Lett.* **52**, 1669.

Von Schultess, C K , Benedek, G. B & de Blois, R. W 1980 Measurement of the cluster size distributions for high functionality antigens Cross-Linked by Antibody. *Macromolecules* **13**, 939.

Weitz, D. A. & Oliveria, M. 1984 Fractal structures formed by kinetic aggregation of aqueous gold colloids. *Phys Rev. Lett.* **52**, 1433

Weitz, D. A. & Huang, J. S. 1984 Self-similar structures and the kinetics of aggregation of gold colloids In *Kinetics of aggregation and gelation* (ed. F Family & D. P. Landau). Amsterdam: North-Holland.

Weitz, D A , Huang, J. S., Lin, M. Y. & Sung, J. 1984 Dynamics of diffusion-limited kinetic aggregation *Phys. Rev. Lett.* **53**, 1651.

Weitz, D. A., Huang, J. S , Lin, M. Y. & Sung, J. 1985a Limits of the fractal dimension for irreversible kinetic aggregation of gold colloids. *Phys. Rev. Lett.* **54**, 1416.

Weitz, D. A., Lin, M. Y., Huang, J. S., Witten, T. A., Sinha, S. K, Gethner, J. S. & Ball, R. C. 1985b Scaling in colloid aggregation. In *Scaling phenomena in disordered systems* (ed. R. Pynn & A Skjeltorp) New York. Plenum.

Weitz, D A. & Lin, M. Y. 1986 Dynamic scaling of cluster-mass distributions in kinetic colloid aggregation *Phys Rev Lett.* **57**, 2037.

Weitz, D A., Lin, M. Y. & Huang, J. S. 1987 Fractals and scaling in kinetic colloid aggregation. In *Physics of complex and supermolecular fluids* (ed. S A. Safran & N. A Clark). New York: Wiley-Interscience.

Wiltzius, P. 1987 Hydrodynamic behavior of fractal aggregates. *Phys Rev Lett.* **58**, 710.

Light-scattering studies of aggregation

By J. G. Rarity[1], R. N. Seabrook[2] and R. J. G. Carr[2]

[1] Royal Signals and Radar Establishment, St Andrews Road,
Great Malvern, Worcestershire WR14 3PS, U.K.

[2] Division of Biotechnology, Public Health Laboratory Service, Centre for Applied
Microbiology and Research, Porton Down, Salisbury, Wiltshire SP4 0JG, U.K.

We discuss dynamic and static light-scattering measurements made during slow (reaction-limited) aggregation of model colloids and immune complex forming proteins. Analysis of the results leads to an understanding of the random aggregates formed in terms of a fractal geometry and measurement of the fractal dimension. Differences in the measured fractal dimensions of the model and protein systems are discussed. The aggregation appears to follow 'Smoluchowski-like' kinetics as measured by a near linear growth of the low-angle light scattering with time. However, the dynamic light-scattering results support a simple power-law model for the aggregate distribution and allow an estimate of this power law to be made.

1. INTRODUCTION

An early theory of the kinetics of aggregation, coagulation and precipitation is that of Smoluchowski (1916). Many studies of aggregation (Reerink & Overbeek 1954; Ottewill & Shaw 1966) have made use of this theory to model the initial stages of the reaction. In recent years there has been an upsurge of interest in the study of aggregation because of the discovery of the fractal nature of the aggregates produced in these reactions (Schaeffer et al. 1984; Weitz & Oliveira 1983). Large-scale computer simulations of aggregation reactions (Meakin 1983; Kolb et al. 1983) have indicated some departure from 'Smoluchowski' kinetics. This has stimulated the development of scaling theories (Kolb 1984; Botet & Jullien 1984; Van Dongen & Ernst 1985) to describe the long-time evolution of the cluster distribution. Two books containing reviews of much of this recent work have appeared (Family & Landau 1984; Stanley & Ostrowski 1986). We use static and dynamic light-scattering techniques at various scattering angles to investigate the applicability of these new theories to the aggregation of model polystyrene colloids and to immune complex formation.

It is well known that low-angle light scattering provides a measure of the mass-average molecular mass of polymers and other macromolecules. Here we exploit this property to obtain a relative measure of the average molecular mass (or aggregation number) of clusters during aggregation. In the early stages of aggregation of model colloids this measure has been shown to increase linearly with time (Lips & Willis 1973; Rarity & Randle 1985) in agreement with the predictions of Smoluchowski (1916). Here we have measured the long-time behaviour of the low-angle scattered intensity and compare the results with recent and past theory.

A fractal object is characterized by a non-integer power-law form for the pair correlation function $g(r)$, which measures the probability of finding material at a point r given material at $r = 0$. For an object embedded in three-dimensional space

$$g(r) - 1 \propto r^{d-3}\, \Psi(r/\zeta), \tag{1}$$

where d is the correlation fractal dimension. Random clusters have a finite extent hence the equation includes a cut-off function Ψ dependent on a measure of cluster radius ζ. Evidence for the fractal nature of random aggregates was originally obtained (Schaeffer et al. 1984; Weitz et al. 1985) from the power-law dependence of scattered intensity $I(Q)$ on the modulus of the scattering vector Q (varied by changing scattering angle) measured when average cluster size is large:

$$I(Q)/P_0(Q) \propto Q^{-d}, \tag{2}$$

where $P_0(Q)$ is the form factor for the seed particles. This relation arises from the power-law form of the pair correlation function in the limit $Q\zeta \gg 1$ (Texeira 1986) and has been shown to hold despite the broad distribution of cluster sizes present in a typical aggregation (Martin 1986).

Another consequence of the fractal nature of the aggregates is that any measure of the radius $R(m)$ of a single cluster is related to the total mass m of the cluster by a power law

$$\frac{m}{m_0} \propto \left(\frac{R(m)}{r_0}\right)^{d_m}, \tag{3}$$

where d_m is a mass fractal dimension, r_0 and m_0 are the radius and mass of individual seed particles and the implied constant of proportionality is of order unity. We expect $d_m = d$ for all random clusters.

Dynamic light scattering at low angles measures an average hydrodynamic radius $\bar{R}_h$ of the suspended clusters. Thus given a narrow cluster-size distribution, the intensity and hydrodynamic radius measured at low angles will be related by a power law equal to a hydrodynamic fractal dimension d_h (Pusey & Rarity 1987):

$$I \propto \bar{R}_h^{d_h}. \tag{4}$$

Real systems are polydisperse and $\bar{R}_h^{d_h}$ involves different moments of the aggregate distribution to the mass average involved in the mean intensity. For the relation to still hold the aggregate distribution should show scale invariance as suggested by the recent theories of aggregation kinetics. Previous work on model colloids (Pusey & Rarity 1987; Wiltzius 1987) supports this conjecture. We expect d and d_h to be equal when hydrodynamic interactions are strong (Meakin et al. 1985).

Immune complex formation (Steensgard 1984) is of wide interest in medical biochemistry. The system is an example of heterocoagulation with limited functionality reactants. As a result we expect some difference from the near universal kinetics and scaling observed in colloidal aggregation. In this work we have measured the hydrodynamic fractal dimension d_h of the immune complexes during the early stages of antigen induced antibody aggregation and the pair correlation fractal dimension d measured in the later stages of the reaction using multiangle static light scattering. We compare the results with similar measurements taken during the salt-induced aggregation of a suspension of polystyrene spheres.

We have also studied the reaction kinetics of the model colloids using low-angle static light scattering coincident with high-angle measurements of the hydrodynamic radius. For large particles and high angles, dynamic light scattering probes the internal motion and rotation of the aggregates. However, the measurement is no longer independent of the form of the aggregate distribution. Scaling theory suggests that the aggregate distribution will have a power-law form with decay exponent τ. It is postulated (Martin & Leyvraz 1986) that in the long time limit the measured hydrodynamic radius will be power law related to the mean aggregate mass. Plotting the high-angle radius against low-angle intensity (a measure of mean aggregate mass) may allow an estimate of τ to be made. We discuss the validity of this measurement in relation to other measures of τ and theoretical predictions (Ball *et al.* 1987).

2. Theory

2.1. *Scaling cluster-size distributions*

The scaling theory of aggregation postulates a cluster distribution of the form

$$\frac{N[m,t]}{N_0(t)} \delta m \sim f[x] \, dx, \tag{5}$$

where $N[m,t]$ is the number of clusters with mass between m and $m + \delta m$ ($\delta m = m_0$) and $N_0(t)$ is the total number of clusters at time t. The $f(x)$ is a scaling function with scaled size

$$x = m(t)/\bar{m}(t) \tag{6}$$

and average cluster mass $\bar{m}(t)$. The constant total mass of the sample sets the normalization of $f(x)$

$$\int x f(x) \, dx = 1. \tag{7}$$

Moments of $N(m,t)$ can be expressed in terms of moments of $f(x)$

$$F_n = \int x^n f(x) \, dx, \tag{8}$$

which remain constant during aggregation. For low-angle light scattering each aggregate scatters as the square of its mass hence the light scattered at time t is given by

$$\frac{I(t)}{I(0)} \propto \frac{\int m^2 N(m,t) \, dm}{\int m N(m,t) \, dm} = \bar{m} F_2. \tag{9}$$

In a similar fashion the hydrodynamic radius measured at low angles is given by (Pusey & Rarity 1987)

$$R_h \propto \bar{m}(t)^{1/d} \frac{F_2}{F_{2-1/d}}. \tag{10}$$

Hence equation (4) holds for polydisperse systems when the assumption of a scaling aggregate distribution holds.

2.2. Aggregation kinetics

The detailed form of the cluster-size distribution is obtained by solution of the coupled equation system

$$\dot{N}(j,t) = \tfrac{1}{2} \sum_{k,l} K(k,l) N(k,t) N(l,t) (\delta_{k+l,j} - \delta_{k,j} - \delta_{l,j}) \tag{11}$$

describing the rate $\dot{N}(j,t)$ of creation and loss of clusters of mass j. The rate constant of formation of $k+l$-mers from clusters of mass k,l is $K(k,l)$. Smoluchowski (1916) postulated a simple rate constant independent of time and k,l and obtained the explicit solution

$$N[j] = (ct)^{j-1}/(1+ct)^{j+1}. \tag{12}$$

The mass-average mass $\bar{m}$ increases linearly with time

$$\bar{m}(t)/m_0 = 1 + 2ct \tag{13}$$

with constant $$2c = (2N_t K_b T)/3\eta, \tag{14}$$

where N_t is the total number of seed particles, η is the solvent viscosity and $K_b T$ is the Boltzmann energy. In the limit of large t and m an approximate scaling result is obtained (Leyvraz 1986) with

$$f(m/\bar{m}) \sim \exp(-2m/\bar{m}). \tag{15}$$

This constant K result is thought to hold for the case of diffusion limited aggregation (DLA) where repulsive barriers are low and clusters stick on impact.

For the reaction limited aggregation (RLA) where substantial repulsive barriers exist between the clusters the probability of sticking on contact will reflect the number of sticking sites available. A possible model for the Kernel is

$$K(k,l) \sim \tfrac{1}{2}(k+l)\,\alpha, \tag{16}$$

where α is a sticking probability ($\alpha^{-1} = W$, the stability factor introduced by Verwey & Overbeek (1948)). The scaling solution valid at long times after initiation of the reaction (Leyvraz 1986) indicates exponential growth of the mass-average aggregate mass

$$\bar{m}(t)/m_0 \sim \exp 2c't, \tag{17}$$

with initial growth rate $2c' = 2c\alpha$. A power-law decaying aggregate distribution

$$f(m/\bar{m}) \sim (m/\bar{m})^{-\tau} g(-m/\bar{m}) \tag{18}$$

is expected with upper cutoff $g(x)$. The decay exponent τ is predicted to be 1.5 (Ball et al. 1987) for a similar mass-dependent Kernel. Electron micrograph studies of gold clusters also indicate a value close to $\tau = 1.5$ (Weitz & Lin 1986).

2.3. *Static and dynamic light scattering at high angle*

High-angle measurements of the apparent radius with dynamic light scattering can be used to estimate τ. Given illuminating light of wavelength λ, scattering angle θ and medium refractive index n the modulus of the scattering vector is given by $Q = 4\pi n \sin\left(\tfrac{1}{2}\theta\right)/\lambda$. The inverse scattering vector is a measure of the lengthscales coherently probed by light-scattering techniques. This lengthscale is shorter at high angles. We can write the total intensity scattered at arbitrary Q as

$$\frac{I(t)}{I(0)} = \frac{\displaystyle\int_{m_0}^{\infty} m^2 N(m,t)\, S(QR)\, dm}{\displaystyle\int_{m_0}^{\infty} mN(m,t)\, dm}, \tag{19}$$

where we include for completeness a lower bound given by the seed particle radius and introduce the cluster structure factor

$$S(QR) \simeq \begin{cases} 1 & QR \ll 1, \\ (QR)^{-d} \sim Q^{-d}m^{-1} & QR \gg 1. \end{cases} \tag{20}$$

The behaviour of equation (19) can be investigated in the limit of high and low cluster masses by choosing a crossover mass $m_1 = aQ^{-d}(QR = 1)$ and by using the limiting values of $S(QR)$ above and below this point. For a power-law polydisperse system with step cutoff function $(g(x) = 0$ for $x > 1$ in equation (18)) we obtain equation (9) in the low Q limit. In the high Q or $\bar{m}$ limit

$$\frac{I(t)}{I(0)} = Q^{-d}\left[1 - \frac{1}{(3-\tau)\,(a^{-1}Q^d\bar{m}(t))^{2-\tau}}\right]. \tag{21}$$

The intensity saturates and scattering becomes dependent on Q^{-d} as in equation (2). The approach to saturation does however give some indication of τ. For a relatively narrow distribution as occurring in the constant kernel kinetics result (equation (15)) $\tau \equiv 0$ hence saturation will be approached rapidly for $\bar{m} > aQ^{-d}$. For $1.5 < \tau < 2$ saturation will be approached much more slowly with $I_{\mathrm{sat}} - I(t)$ having a power-law form.

The apparent measured radius in a dynamic light-scattering experiment can be calculated in the same way from

$$R_{\mathrm{app}}(t) \propto Q^2 \frac{\displaystyle\int_{m_0}^{\infty} m^2 N(m,t)\, S(QR)\, dm}{\displaystyle\int_{m_0}^{\infty} m^2 N(m,t)\, S(QR)\, \Gamma(QR)\, dm}, \tag{22}$$

where the linewidth or decay constant for a single cluster is given by (Martin & Leyvraz 1986)

$$\Gamma(QR) = \begin{cases} D(m)\, Q^2 \sim Q^2 m^{-1/d} & m \ll m_1, \\ Q^2 m^{-1/d} h(QR) \sim Q^{2+\omega} m^{(\omega-1)/d} & m \gg m_1, \end{cases} \tag{23}$$

where $0 \leqslant \omega \leqslant 1$. The high Q form arises from consideration of dynamic scattering from a single large aggregate. When the cluster is rigid the intensity fluctuations arise from rotation. The mean square angle of rotation at time t is given by

$$\langle \Delta\phi^2 \rangle = D_R t \tag{24}$$

and the typical intensity fluctuation time $t_c = \Gamma^{-1}$ is given by the mean time taken for particles at typical radius R to diffuse a distance Q^{-1} hence

$$\Gamma = t_c^{-1} \sim D_R Q^2 R^2 \sim Q^2 m^{-1/d} \tag{25}$$

given spheroidal particles where $D_R \sim R^{-3}$. This result corresponds to $\omega = 0$ in equation (23). A similar argument for completely flexible particles leads to the well-known polymer result (De Gennes 1979) $\Gamma \sim Q^3 m^0$ corresponding to $\omega = 1$.

Substituting equation (23) into equation (22) and doing the integrations above and below the crossover m_1 we obtain in the limit of large $\bar{m}$

$$R_{app} \propto \begin{cases} Q^{2-d} & 3 > \tau > 2, \\ \bar{m}^{(2-\tau)}Q^{-1+d(2-\tau)}[1+f_a(\bar{m}/m_1)] & 2 > \tau > 2-(1-\omega)/d, \\ Q^{-\omega}\bar{m}^{(1-\omega)/d} & \tau < 2-(1-\omega)/d, \end{cases} \tag{26}$$

where the approach to asymptotic behaviour in the intermediate τ region has the form

$$f_a(\bar{m}/m_1) = \frac{(\bar{m}/m_1)^{-(2-\tau)}}{(3-\tau)} - \frac{(3-\tau-1/d)(\bar{m}/m_1)^{2-\tau-(1-\omega)/d}}{1-\omega/d}. \tag{27}$$

For RLA (slow aggregation as studied here) we expect $d \sim 2$ implying reasonably rigid particles and hence $\omega = 0$. For τ between 1.55 and 1.95 (the solution is not strictly valid at the crossover points) we should see a power-law dependence of the measured radius on measured molecular mass (or low-angle intensity)

$$I(t)_{\theta \to 0} \propto \bar{R}_{app}^{1/(2-\tau)}. \tag{28}$$

This coincides with a power-law dependence of R_{app} on scattering vector

$$R_{app} \propto Q^{d(2-\tau)-1} \tag{29}$$

as proposed by Martin & Leyvraz (1986). The approach to the asymptotic behaviour appears to be slow but there is cancellation of terms as indicated by the simplistic form shown in equation (27). A test of the proximity to asymptotic behaviour can be obtained by measuring the saturation (equation (21)) of intensity at high angle while monitoring the low-angle intensity. The exact form of the approach to saturation will depend on the exact behaviour of scattering factors in the crossover region $m \sim m_1$.

3. EXPERIMENTS AND RESULTS

3.1. *Polystyrene model colloids*

Polystyrene spheres of nominal radius 22 nm were suspended in pure water at concentrations around 10^{12} particles per millilitre. Before all experiments the samples were repeatedly filtered through 0.22 μm pore filters. Aggregation was induced by introducing salt solution to a concentration of *ca.* 0.15 M NaCl through a 0.45 μm filter and rapid mixing. For these conditions we expect a slow aggregation about 10^{-3} times the diffusion limited rate. Low-angle measurements of hydrodynamic radius and intensity confirm the theory of equation (4). The measured hydrodynamic dimension d_h appears to be equal within experimental error to the correlation dimension d measured after saturation from a logarithmic plot of intensity against scattering vector. We measured (Pusey & Rarity 1987)

$$d_h \approx d = 2.08 \pm 0.05. \tag{30}$$

This value is typical of slow aggregation.

In the present series of experiments on this model colloid system the intensity at low angle ($\theta = 12.8°$) was measured coincident with the high angle ($\theta = 90°$) apparent hydrodynamic radius R_{app}. Measurements were made every few minutes after aggregation. Standard photon correlation equipment was used and the hydrodynamic radius was obtained from an average of second- and third-order cumulant fits to the measured correlation function (Koppel 1972). A Krypton ion laser operating at 647 nm wavelength was used as illuminating source. A logarithmic plot of low-angle intensity against time is shown in figure 1. No

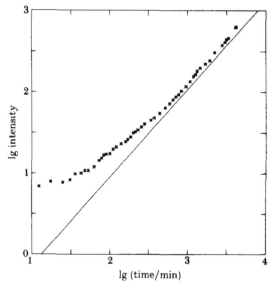

FIGURE 1. Doubly logarithmic plot of low-angle scattered intensity from aggregating polystyrene spheres as a function of time in minutes; $\theta = 12.8°$ and illuminating wavelength is 647 nm Solid line shows a gradient of 1.08

saturation effects can be seen at later times and there appears to be near linear growth of intensity with time. We measure

$$I(t) \propto t^{1\ 08 \pm 0.05} \tag{31}$$

in the later stages of the reaction. This would be expected if the reaction were near diffusion limited or constant kernel kinetics applied. For comparison, figure 2

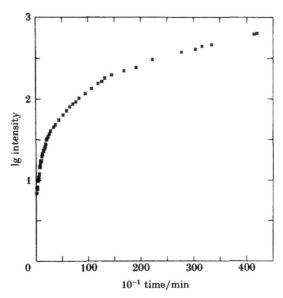

FIGURE 2. Logarithmic-linear plot of the data in figure 1.

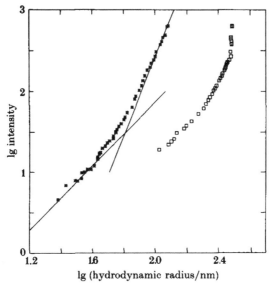

FIGURE 3. Doubly logarithmic plot of low-angle scattered intensity as a function of apparent hydrodynamic radius R_{app} measured at 90° for the polystyrene system (∗). The linear region in the later stages of the reaction shows a gradient of 5 (upper solid line) whereas in the early stages the limiting gradient appears to be the expected value of 2 (lower solid line). Also shown is the saturation of the high-angle scattered intensity (□) displayed in suitably scaled units.

shows the same data plotted logarithmic-linearly. Clearly exponential behaviour can only be inferred in the early and later stages of the reaction.

Figure 3 shows a logarithmic plot of low-angle intensity as a function of the measured high-angle hydrodynamic radius. In the early stages of the reaction we see an approximate R_h^2 dependence confirming a fractal dimension d_h near two. In the region $QR_h \simeq 1$ we see a crossover to a higher power-law dependence. A visual fit to the data shown gives

$$I(t) \propto R_{app}^{5 \pm 0.5}. \tag{32}$$

If we are in the true asymptotic region and a power-law size distribution holds this implies (equation (28)) that

$$\tau = 1.8 \pm 0.05. \tag{33}$$

However, this result could imply some scaling behaviour of the internal fluctuations of the aggregates with a value of $\omega \simeq 0.6$ (equation (26), $\tau < 2 - (1 - \omega)/d$). The existence of a non-integer value of ω is unlikely. Both explanations imply a $Q^{-0.6}$ dependence of R_{app}. This agrees with other measurements on silica aggregates (Martin 1987; Martin & Schaeffer 1984) where a non-integer power-law dependence of apparent radius on scattering vector $Q^{-0.7}$ was measured. Preliminary measurements of R_{app} as a function of Q on the polystyrene system also show a $Q^{-0.7}$ dependence. A check of closeness to the asymptotic limit is provided by a logarithmic plot of low-angle intensity against high-angle intensity also shown in figure 3. Clearly the high-angle intensity is well into the saturation régime when the power law behaviour is observed.

3.2. *Antibody–antigen aggregation*

To 1 ml phosphate-buffered saline (PBS) was added 84 µl of antisera (5 mg ml^{-1} anti-human IgG, Sigma Chemical Co.). This was filtered several times through a 0.22 µm filter. Then 642 µl of the solution were transferred to a sample cuvette and studied by dynamic light scattering at 30°. The cleanliness of the sample was assured when a hydrodynamic radius less than 10 nm was measured, indicating that the proteins in the cuvette were mainly present in unaggregated form. The aggregation was started by addition of 42 µl of 0.22 µm filtered purified human IgG (200 µg ml^{-1}) (Sigma Chemical Co.). This is equivalent to 12 µg per millilitre of human IgG in the cuvette. The mixing was performed by rapid pipette action. Repeated measurements of low-angle hydrodynamic radius and intensity were made starting a few seconds after mixing. For this work an Ar ion laser operating at 514 nm wavelength was used and the scattering angle was 30°. Although this corresponds to a larger value of Q than the earlier polystyrene studies the smaller size of the original particles ensures a reasonable region of growth that can be studied before the low-angle approximation $QR < 1$ breaks down. The aggregation rate could be enhanced by addition of polyethylene glycol (PEG) of mean molecular mass 6000 (PEG 6000, BDH Chemical Co.) to the PBS buffer. Enhancement in initial reaction rate of a factor of about 10 could be seen with a PEG concentration of 1 % (by volume). Results showing power-law dependence of the low-angle hydrodynamic radius on low-angle intensity for three concentrations of PEG (0 %, 0.5 % and 1.0 %) are shown in figure 4. The results overlay extremely

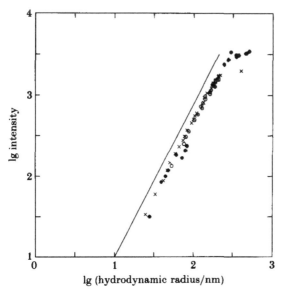

FIGURE 4. Doubly logarithmic plot of low-angle scattered intensity against hydrodynamic radius for the early stages of antibody–antigen aggregation. Symbols: ×, no added PEG; ⊙, 0.5% PEG; ✦, 1% PEG. Solid line shows a gradient of 1.9.

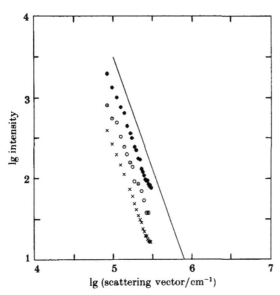

FIGURE 5. Doubly logarithmic plot of scattered intensity as a function of scattering vector measured in the later stages of antibody–antigen aggregation. Symbols: ×, no added PEG; ⊙, 0.5% PEG; ✦, 1% PEG. Solid line shows a gradient of 2.65.

well reflecting the repeatability of starting conditions and similar fractal dimension for aggregates formed. The estimated power law is

$$I(t) \propto R_h^{1.9 \pm 0.1}, \tag{34}$$

which after correction for the non-zero angle (Pusey & Rarity 1987) indicates a hydrodynamic fractal dimension close to $d_h = 2.1$ as expected from a slow aggregation. After saturation of the low-angle intensity further measurements of scattered intensity as a function of scattering angle were made. Typical logarithmic plots of intensity against Q are shown in figure 5. Clearly the gradient is much greater than the expected value for $d = 2.1$. We measure

$$I(Q) = Q^{-2.65 \pm 0.05} \tag{35}$$

from several independent measurements taken after saturation. This implies $d = 2.65 \pm 0.05$. There appears again to be no systematic variation of the slope for the different PEG concentrations. Deviation from straight-line behaviour at high Q values (scattering angles larger than 120°) is caused by background light arising from partial reflections at the cuvette walls of some of the strong forward scattering. We have also measured the low-angle scattered intensity as a function of time and obtain near linear growth with time as seen in the model colloid experiments.

4. DISCUSSION AND CONCLUSIONS

We have measured the long time growth of mean cluster size using low-angle light-scattering data from slow reaction limited aggregation of polystyrene colloids. The mean cluster size appears to grow linearly with time over two orders of magnitude. Regions where short sections of data may be confused with exponential growth are identified in the early stages and later stages of the reaction. As the initial rate of growth is only a factor of 1000 down on the diffusion limited rate the conditions where a mass-dependent Kernel strongly affects the growth of the mean mass may never be reached. However, linear growth also appears to occur in the antibody–antigen aggregation data reported here and in heat-induced aggregation of proteins (Feder *et al.* 1984) where the sticking probability is extremely small ($\alpha = 10^{-9}$).

The dynamic light-scattering data at high angle do, however, support the hypothesis of a power-law decay of the cluster-size distribution. The power-law growth of R_{app} indicates a cluster-mass distribution of the form

$$N(m) \sim m^{-1.80 \pm 0.05} g(-m/\bar{m}). \tag{36}$$

The main difficulty in testing the validity of this type of result lies in ensuring that the true asymptotic gradient has been reached. In this work we have shown that this condition is satisfied when the high Q scattering saturates allowing a reliable estimate of asymptotic gradients to be obtained. The detailed approach to saturation of high Q intensity measurements may also provide a useful indicator of the form of the cluster mass distribution. The theoretical prediction of $\tau = 1.5$ for reaction limited aggregation is a little lower than our measured result. This may be accounted for by the non-universality of the kinetics that may be in the crossover region between diffusion limited and reaction limited; however, as these results were taken over a period of three days the size distribution could be affected by more rapid sedimentation of the larger clusters.

Measurements of antigen-induced aggregation of antibody indicate a hydro-dynamic fractal dimension of $d_h = 2.1 \pm 0.1$ in the early stages of the reaction whereas multiangle measurements made after saturation of the low-angle intensity indicate a correlation fractal dimension $d = 2.65 \pm 0.05$. This does not indicate any measurement-dependent behaviour of the fractal dimension and we expect $d_h = d$ for well-behaved mass fractals. However, we believe that this difference in measured dimension arises from a restructuring of the aggregates with time because of the reversibility of the individual antibody–antigen bonds (Smith & Skubitz 1975). There is also some evidence for antibody–antibody bonding within the aggregates which eventually leads to precipitation of large dense complexes a long time (ca. 24 h) after the start of aggregation (Moller & Steensgard 1979). There may also be some non-specific aggregation of non-immune proteins with the clusters as the antisera is not pure antibody. The high fractal dimension does seem to be characteristic of protein aggregation. Similar fractal dimensions have been measured from aggregation of casein micelles (Horne 1987) and heat aggregation of immunoglobulins (Feder et al. 1984). It is interesting to note that the antibody–antigen reaction is a heterocoagulation where the mean field theory of Smoluchowski may break down. This could lead to concentration inhomogeneities that affect scattering measurements. One might expect to see such behaviour with high concentrations of added polymer (PEG). Our results indicate a near universal behaviour independent of polymer concentration within the limited range of concentrations studied.

REFERENCES

Ball, R. C., Weitz, D. A., Witten, T. A. & Levraz, F. 1987 *Phys. Rev. Lett.* **58**, 274.
Botet, R. & Jullien, R. 1984 *J. Phys.* A **17**, 2517.
De Gennes, P.-G. 1979 *Scaling concepts in polymer physics.* Ithaca and London: Cornell University Press.
Family, F. & Landau, D. P. (ed.) 1984 *Kinetics of aggregation and gelation.* New York and Amsterdam. North Holland.
Feder, J., Jossgang, T. & Rosenqvist, E. 1984 *Phys. Rev. Lett.* **53**, 1403.
Horne, D. S. 1987 *Faraday Discuss. chem. Soc.* **83**, 259.
Kolb, M. 1984 *Phys. Rev. Lett.* **53**, 1654.
Kolb, M., Botet, R. & Jullien. R. 1983 *Phys. Rev. Lett.* **51**, 1123.
Koppel, D. E. 1972 *J. chem. Phys.* **57**, 4814.
Leyvraz, F. 1986 In *On growth and form* (ed. H. E. Stanley & N. Ostrowski), p. 136. Boston. Martinus Nijhoff.
Lips, A. & Willis, E. 1973 *J. chem. Soc. Faraday Trans. I.* **69**, 1226.
Martin, J. E. 1986 *J. appl. Crystallogr.* **19**, 25.
Martin, J. E. 1987 *Phys. Rev.* A **36**, 3415.
Martin, J. E. & Leyvraz, F. 1986 *Phys. Rev.* A **34**, 2346.
Martin, J. E. & Schaeffer, D. W. 1984 *Phys. Rev. Lett* **53**, 2457.
Meakin, P. 1983 *Phys. Rev. Lett* **51**, 1119.
Meakin, P., Vicsek, T. & Family, F. 1985 *Phys. Rev.* B **31**, 564.

Moller, N. P H & Steensgard, J. 1979 *Immunology* **38**, 641.
Ottewill, R. H. & Shaw, J. N 1966 *Discuss Faraday Soc.* **42**, 143.
Pusey, P N. & Rarity, J. G. 1987 *Molec. Phys.* **62**, 411.
Rarity, J. G. & Randle, K. J. 1985 *J. chem. Soc Faraday Trans. I.* **81**, 285.
Reerink, H & Overbeek, J. Th. G. 1954 *Discuss. Faraday Soc* **18**, 74
Schaeffer, D., Martin, J. E., Wiltzius, P. & Cannel, D. S. 1984 *Phys Rev. Lett.* **52**, 2371.
Smith, T. W. & Skubitz, K M. 1975 *Biochemistry* **14**, 1496.
Stanley, H. E. & Ostrowski, N. (ed.) 1986 *On growth and form.* Boston: Martinus Nijhoff.
Steensgard, J. 1984 *Immunol. Today* **5**, 7.
Texeira, J. 1986 In *On growth and form* (ed H. E. Stanley & N. Ostrowski), p. 145. Boston: Martinus Nijhoff.
Van Dongen, P. G. J. & Ernst, M H. 1985 *Phys. Rev. Lett.* **54**, 1396.
Verwey, E. J. W. & Overbeek, J. Th. G. 1948 *Theory of the stability of lyophobic colloids.* Amsterdam. Elsevier.
Von Smoluchowski, M. 1916 *Phys Z.* **17**, 585
Weitz, D A & Oliviera, M. 1983 *Phys. Rev Lett.* **52**, 1433.
Weitz, D. A , Huang, J S, Lin, M Y & Sung, J. 1985 *Phys. Rev. Lett* **54**, 1416.
Weitz, D. A. & Lin, M. Y 1986 *Phys. Rev. Lett.* **57**, 2037.
Wiltzius, P. 1987 *Phys. Rev. Lett.* **58**, 710

Discussion

D. A. WEITZ (*Exxon Research and Engineering Co., Annandale, U.S.A.*). The experimental data that we have presented in our paper for $\bar{D}_{eff} \sim R_{app}^{-1}$ for the case of reaction-limited colloid aggregation are similar to those presented in this paper by Dr Rarity. However, we differ as to interpretation.

Dr Rarity finds

$$I(0) \sim R_{app}^{1/2-\tau}.$$

This analysis is based on the assumption that $\tau > 2 - 1/d_f$ so that R_{app} is no longer sensitive to the diffusion coefficients of the larger clusters. Although this may be true in the asymptotic limit, unrealistically large cluster masses, of order 10^{20}, would have to be achieved for τ close to 1.5.

In our data, we observe

$$\bar{D}_{eff}/\bar{D} = (q\bar{R})^\alpha$$

where, for $\tau = 1.5$, we find $\alpha \approx 0.5$ if we force a power-law interpretation around $1 < q\bar{R} < 10$, the region that corresponds to Dr Rarity's data. To compare with Dr Rarity's data,

$$I(0) \sim R_{true}^{d_f} \sim R_{app}^{d_f/(1-\alpha)},$$

which now involves the fractal dimension of the clusters. This predicts an exponent of 4.2.

An additional feature that may have to be considered is the effect of differential sedimentation due to gravity for the very polydisperse cluster mass distribution obtained for RLCA. We find that it significantly changes the shape of the master curve, as shown in figure D1, which is obtained by using gold colloids without inverting our sample cell every 15 min to eliminate gravitational sedimentation. The slope of the data is somewhat steeper, as can be seen by comparing it to the

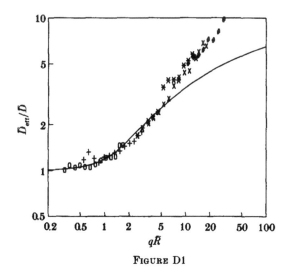

theoretical prediction for $\tau = 1.5$, shown by the solid line in the figure. The sedimentation was clearly visible in the sample cell and occurred in the space of several hours. Although gold is substantially denser, sedimentation will also occur in polystyrene, but on a timescale of a few days, consistent with the duration of Dr Rarity's experiment.

Time-series analysis

By D. S. Broomhead and R. Jones

Royal Signals and Radar Establishment, St Andrews Road,
Great Malvern, Worcestershire WR14 3PS, U.K.

The fractals of interest in this paper are sets of states attainable by a system whose dynamical behaviour is governed by nonlinear evolution equations. Experimentally such sets, normally associated with chaotic dynamics, can occur in chemical reaction systems, fluid flows, electronic oscillators, driven neurons: the list has been growing rapidly over the past 10–15 years. Experimental studies of these systems generally produce time series of measurements that have no obvious fractal properties. Classical signal-processing methods provide little help. However, recently, techniques have been developed that are capable of extracting geometrical information from time-series data. This paper will review some of these methods and their application to the study of the geometry of invariant sets underlying the dynamics and bifurcations of some experimental and model systems.

1. Introduction

Mass action kinetics give rise to nonlinear evolution equations for systems of chemical reactions; similarly, fluid flows are governed by nonlinear laws in all but special limiting cases. Even stock markets, it seems, when controlled by computer program, evolve as deterministic nonlinear dynamical systems. The realization that nonlinearities have important consequences on the dynamics of real systems is not new (except, perhaps, in the case of the stockmarket). Over the past century (Poincaré 1899), and particularly over the past two decades (see, for example, Guckenheimer & Holmes 1983; Bergé *et al.* 1984) there has been a strong interest in the mathematics of nonlinear dynamical systems. What is relatively new is the attempt to relate the geometric but abstract mathematical understanding of nonlinear systems to experimental observations (for example, Meyer-Kress 1986). One approach to this, put in the context of fractal geometries, will be reviewed in the following.

2. An example of a fractal attractor

Fractals occur in many ways in dynamical systems (Fischer & Smith 1985; Peitgen & Richter 1986). Here the specific case of fractal attractors in continuous time experimental systems will be considered. The techniques we shall use are applicable whenever we have time-series data taken from a system whose dynamics are described by a finite dimensional attractor, be it fractal or not. The specific example we shall consider consists of a time series of voltage measurements taken from an electronic nonlinear oscillator. Data were collected by using a 12 bit analogue to digital (a/d) converter sampling at a frequency of 10 kHz, which provided about 100 data samples in one cycle of the basic frequency of the oscillator. A short sequence of the data is shown in figure 1. In the rest of this section we shall attempt to show: (a) that despite their rather smooth appearance, these data imply the existence of an underlying fractal attractor; and (b) the quality of the data is not sufficient for the fractal to be observed directly.

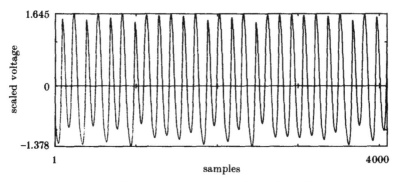

FIGURE 1. Time-series data from an electronic oscillator.

Figure 2 shows the same data as figure 1, this time however, plotted as a trajectory in the same way as one might represent the solution to a set of coupled differential equations. We think of this trajectory as a curve drawn in a three-dimensional space and finally projected onto a plane. The details of this process are postponed until §§ 3 and 4. The small rectangles around the border show cross

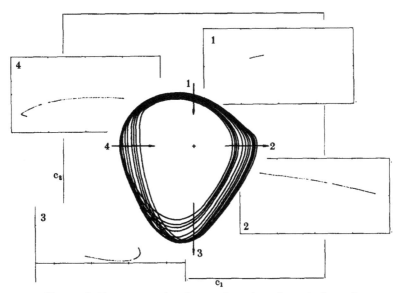

FIGURE 2. Reconstructed trajectory from data shown in figure 1, and cross sections taken at the places indicated.

sections taken through the three-dimensional object at the places marked. These may be used to build up a picture of the full three-dimensional motion of the flow.

Consider a sequence of sections taken in the order in which the flow transforms one section into the next. The result of following the flow for one complete circuit is that the pieces of the trajectory starting in section 1 are first stretched apart and then folded back on themselves. They thus return to section 1 within the same straight line segment in which they began. One can see from the projection of the flow how this is accomplished: the trajectories spiral outwards until beyond a certain radius, they are reinjected nearer to the centre of the spiral. This repeated stretching and folding process gives rise, not only to chaotic dynamics, but also to a thin fractal attractor. This can be seen from the idealized drawing in figure 3, which shows three iterations of a stretching and folding map applied to a rectangle ABCD. The connection with figure 2 is made by assuming that the straight line segment in section 1 is actually a very thin rectangle. The action of the flow is then to repeatedly stretch the rectangle horizontally, fold it and then compress it strongly in the vertical direction. Figure 3 shows that the result of repeating this process is a fractal that is a very long, thin multiply folded image of the original rectangle. This phenomenon was noted by Rössler (1977) who compared it with making flaky pastry, which, assuming a conscientious chef, may be thought of as a culinary fractal. Roux *et al.* (1983) observed the same qualitative behaviour in a chemical reaction system.

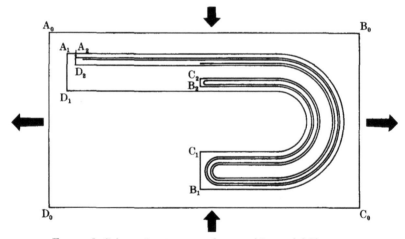

Figure 3. Schematic operation of a stretching and folding map

Having identified a fractal-generating mechanism in our data we would expect that, by enlarging the sections shown in figure 2, it is possible to resolve some of the detailed structure. However, when this procedure is done it appears that the dominant contribution to the thickness of the sections is noise on the data. This statement will be made more precise later; here we remark that this situation is not unusual. In common with many naturally occurring fractals, the structure of the chaotic attractor is produced by a balance between an unstable growth, in this case the stretching process, and a stable, limiting process that takes place on a different scale. The limiting process here is the compressive folding of the attractor provided by the nonlinearity. Thus it is common for a chaotic attractor to have the appearance of a multisheeted structure when sectioned in the stable direction. A consequence of this is that even the largest scales of such structures are small and therefore difficult to extract from experimental noise.

The process we have observed in this experimental system is well known in theoretical studies, for example Tomita & Kai (1978) working on a forced chemical oscillator and more recently Thompson (1989) who studied a forced mechanical oscillator. Both these systems generate sequences of sections similar to figure 2. In autonomous systems we have already mentioned Rössler's work and the behaviour he described has been related to the existence of a type of homoclinic orbit (Gaspard & Nicolis 1983). That is, a trajectory that spirals out from a fixed point until it reaches a critical radius and is then reinjected exactly onto the fixed point. We know that by small adjustments in the control parameters of the oscillator, from those used to generate the data shown in figure 1, it is possible to find states where homoclinic orbits exist. There is a considerable body of theory that describes the dynamics and fractal structures associated with homoclinic orbits (Shilnikov 1965; Glendinning & Sparrow 1984; Gaspard et al. 1984). Other physical systems for which there is evidence for homoclinic orbits and the corresponding folding and stretching processes include chemical reaction systems (Argoul et al. 1987) and the Taylor–Couette fluid flow experiment (T. Mullin, personal communication 1988).

3. PHASE PORTRAIT RECONSTRUCTION

In this section we describe how a sampled data sequence such as in figure 1 may be used to construct a multidimensional trajectory as in figure 2. The properties of this construction will be discussed in the following section.

The particular method we use is founded, conceptually, on the use of a delay register. Data are fed into the register and propagate sequentially through until they are lost from the other end n clock cycles later. At any instant the register contains a sequence of n consecutive data values, say, $(v_i, v_{i+1}, ..., v_{i+n-1})$. One clock pulse later after a new datum has entered the register its contents are: $(v_{i+1}, v_{i+2}, ..., v_{i+n})$. Each sequence of n data values can be thought of as an n-dimensional vector (a column vector by convention) $x_i = (v_i, v_{i+1}, ..., v_{i+n})^T$, whereas the sequence of n-vectors generated by clocking the data through the delay register can be thought of as a discrete trajectory in an n-dimensional euclidean space. This space will be referred to as the 'embedding space' for reasons to be made clear in the next section.

By using the concept of the delay register a time series of experimental measurements can be converted into a set of points in the embedding space. We can now study the static geometry of this set by considering it to have been generated by sampling a distribution function or measure that is invariant under the time evolution of the system (Farmer et al. 1983; Eckmann & Ruelle 1985; Meyer-Kress 1986). It is also possible to study the dynamics of the set by looking at the way in which points map onto each other under the action of the flow. We shall be concerned primarily with characterizing the geometry of the invariant measure.

The simplest property of the measure is its mean, the n-dimensional vector $\langle v \rangle \delta$ where $\langle v \rangle$ is the mean value of the time series and $\delta = (1, 1, ..., 1)^T/n^{\frac{1}{2}}$ is the unit vector on the diagonal of the embedding space. Here we shall use coordinates for the embedding space that are centred on the mean by making the replacement $x_i \rightarrow x_i - \langle v \rangle \delta$. The second moments of the measure centred at the mean are found by diagonalizing the covariance matrix:

$$\frac{1}{N} \sum_{i=1}^{N} x_i x_i^T. \tag{1}$$

This procedure is known as principle component analysis, or the derivation of a Karhunen–Loeve basis (Devijver & Kittler 1982). It is related to the singular value decomposition (Golub & van Loan 1985; Pike *et al.* 1984) of the $N \times n$, trajectory matrix, X, whose rows consist of the row vectors x_i^T. The covariance matrix is actually X^TX, therefore the right singular vectors of X are the eigenvectors of the covariance matrix whereas the corresponding singular values of X are the positive square roots of the eigenvalues of the covariance matrix.

There is a physical analogy to this procedure. If we think of the measure as the local density or mass distribution of a body in the embedding space, finding the singular vectors (principal basis) corresponds to finding the axes of inertia of the body. The singular values are then the root-mean-square projections of its mass distribution onto each of the axes of inertia. In the simple case where the data sequence is a noise process consisting of independently, identically distributed samples it is easy to show that the x_i are distributed isotropically. The singular vectors may be thought of as the axes of inertia of an n-dimensional sphere and the singular values are all equal and measure the width of the distribution of the data values.

Mathematically, the singular vectors are an orthonormal basis for the embedding space, which has the property that it spans a nested set of optimal subspaces. Thus, for any $k \leqslant n$, if we seek to expand the vectors in the delay register as a linear combination of k-orthonormal n-vectors then the least mean-square error, taking the average over all the data, is obtained by using the k-singular vectors corresponding to the k largest singular values. If we assume the isotropic noise process is superimposed on a signal, then the first k singular vectors span the k-dimensional subspace of the embedding space for which the signal-to-noise ratio is maximum.

These points are illustrated in figure 4, which shows the spectrum of singular values and the corresponding singular vectors for the time series shown in figure 1. Plotted on the singular spectrum is an estimate of the quantization noise obtained by comparing the standard deviation of the data with the resolution of the A/D converter. In fact we find a flat noise floor (with corresponding noisy singular vectors) with a magnitude of about one third of this estimate (the discrepancy is accounted for by noting that the standard deviation underestimates the range of the A/D converter covered by the signal). This kind of singular spectrum is

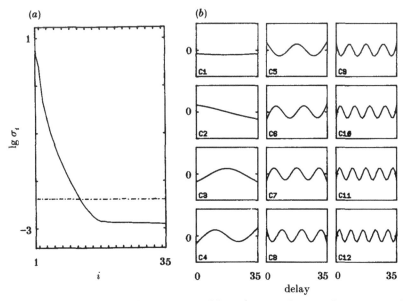

FIGURE 4. (a) Singular value spectrum of data shown in figure 1; dash-dot line shows estimate of quantization noise. (b) Normalized singular vectors.

interpreted to mean that the data consists of samples of the desired signal from the oscillator together with a superimposed uncorrelated noise process (usually dominated by quantization error). Consequently the embedding space can be decomposed into two parts:

(i) the subspace spanned by the first 16 or 17 singular vectors where the signal-to-noise ratio is greater than unity;

(ii) the orthogonal complement spanned by the remaining singular vectors where the dominant process is the experimental noise.

In general we set a minimum level of signal to noise and compare this with the singular spectrum. There will be d, where $d \leqslant n$, singular values above this minimum level. The effective embedding space is then not $\mathbb{R}^n$, but $\mathbb{R}^d$, the subspace spanned by the first d singular vectors (Broomhead & King 1986a, b; King et al. 1987). Thus we use the projected outputs from the delay register:

$$\hat{x}_i = \sum_{j=1}^{d} (c_j^T \cdot x_i) c_j, \tag{2}$$

where the singular vectors are written c_j and

$$c_j^T \cdot x_i = \sum_{k=1}^{n} c_{jk} v_{i+k-1}. \tag{3}$$

In particular, the trajectory in figure 2 was obtained by plotting the sequence of points $(c_1^T \cdot x_i, c_2^T \cdot x_i)$. From the form of the singular vectors shown in figure 4, the control of noise is obtained by introducing averaging over the contents of the delay register.

4. Embeddings and equivalent dynamics

The state space of the oscillator in the previous sections is three-dimensional because application of Kirchhoff's Laws to the circuit tells us that three parameters are needed to specify uniquely its state at any time. In practise there exist within the state space a number of attracting sets that capture almost all the possible initial states of the oscillator. Thus, the oscillator ultimately evolves in a rather restricted manner, exploring a lower-dimensional subset of its state space. This is even true of many fluid systems for which the state spaces are infinite dimensional (Constantin *et al.* 1985). The data-processing technique described in the previous section is based on this observation.

The idea of using sequences of measurements to gain geometric information was first proposed independently by Packard *et al.* (1980) and Takens (1981), the basic assumption being that the attracting set, which may as we have seen, be a fractal, is a subset of an attracting manifold. For our present purposes we think of an m-dimensional manifold as a nonlinear space that may be approximated locally by m-dimensional linear spaces: its tangent spaces. For example, the surface of a sphere is a 2-manifold, which may be approximated locally by tangent planes. There is a classical result due to Whitney (Hirsch 1976) that states that, under certain general conditions, it is possible to embed an m-manifold in an n-dimensional euclidean space if $n \geqslant 2m+1$. By 'embed' we mean that there is a mapping say Φ, from the manifold to the euclidean space such that the manifold and its image have the same topology. For points in the image this mapping must have a unique inverse and both the mapping and its inverse must be continuously differentiable. The reason for introducing this technicality concerns the equivalence of dynamical systems defined on a manifold. If such a relation, Φ, exists and is used to map a dynamical system to another manifold so that the image of an orbit $y(t)$ on one manifold is an orbit $x(t) = \Phi(y(t))$ on the other, then there is a very close relationship (termed differential equivalence (Guckenheimer & Holmes 1983)) between the two dynamical systems. In particular, limit sets are mapped into one another by Φ and Φ^{-1}, therefore fixed points, limit cycles and chaotic attractors on one manifold correspond to fixed points, limit cycles and chaotic attractors on the other. Moreover, the stability of those sets is preserved and indeed, because of the smoothness, so are ratios of eigenvalues and characteristic exponents. Of direct interest in the study of fractal limits sets is the fact that the fractal dimensions of the sets are preserved (even if Φ and Φ^{-1} are Lipschitz continuous (Arneodo & Holschneider 1987)).

The reason for this apparent digression is that the delay register concept has been proved by Takens (1981) to be an embedding if the dynamics being observed correspond to motion on an m-manifold and if the dimension of the embedding space is large enough. The condition on the dimension of the embedding space is Whitney's condition stated above. For noiseless data this is $n \geqslant 2m+1$, where n is the number of stages in the delay register. However, as we have shown in §3 the presence of experimental noise means that only a d-dimensional subspace of the embedding space is explored, thus the condition, for noisy data, is

$$d \geqslant 2m+1, \tag{4}$$

where $d = \text{rank}\,(X)$. Of course the value of m is not known *a priori*; however, this too can be estimated by using a singular value technique as we shall show.

5. DIMENSIONS AND MEASURES

The embedding procedure described in the previous sections generates, in principle, an image of the attractor in the embedding space. In practice a finite set of points $\{\hat{x}_i\,|\,i = 1, ..., N\}$ is obtained that may be thought of as being sampled from the natural measure induced on the image of the attractor. Formally the density of this measure is

$$d\mu(\hat{x}) = \lim_{N\to\infty} \frac{1}{N}\sum_{i=1}^{N} \delta(\hat{x} - \hat{x}_i)\,d\hat{x}, \tag{5}$$

which, for finite amounts of data must be approximated as

$$d\mu(\hat{x}) \approx \frac{1}{N}\sum_{i=1}^{N} \delta(\hat{x} - \hat{x}_i)\,dx. \tag{6}$$

This section gives a brief summary of various fractal dimensions that may be used to characterize the measure μ. These quantities are invariant under the embedding process and therefore give direct information about the physical system from which the data were obtained.

(a) *Pointwise dimension*

The pointwise dimension, d_p, (Young 1982; Farmer *et al.* 1983) is obtained by studying the scaling behaviour of coarse-grained estimates of the density of points on the attractor. Consider the open ball $B_e(\hat{x}) = \{\hat{y} \in \mathbb{R}^d\,|\,|\hat{x} - \hat{y}| < \epsilon\}$, the set of all points within a radius ϵ of the chosen centre $\hat{x}$. The measure of this ball $\mu[B_e(\hat{x})]$ is

$$\mu[B_e(\hat{x})] = \int_{B_e(\hat{x})} d\mu(\hat{x}'), \tag{7}$$

that is, from equation (5) or equation (6), the fraction of the total number of the $\{\hat{x}_i\}$ that are found within the ball. In essence, $\mu[B_e(\hat{x})]$ is an estimate of the local density of points using a length scale $ca.$ ϵ. The local scaling of this quantity with length is then given by the local exponent $d_p(\hat{x})$:

$$d_p(\hat{x}) = \lim_{\epsilon\to0} \frac{\ln\,[\mu[B_e(\hat{x})]]}{\ln \epsilon}. \tag{8}$$

The pointwise dimension is defined if $d_p(\hat{x})$ is independent of $\hat{x}$ for almost all choices of $\hat{x}$ (with respect to μ):

$$d_p(\hat{x}) = d_p, \text{ a constant for almost all } \hat{x} \text{ with respect to } \mu$$

$$\Rightarrow \text{the pointwise dimension} = d_p.$$

It can be shown (Young 1982) that if the pointwise dimension exists then it equals the Hausdorf dimension of the measure, μ (if we look at the Hausdorf dimension of all sets of full measure then the Hausdorf dimension of the measure is the infimum of these). However, there is no reason why chaotic attractors should not be multifractal objects, in which case the local exponent can be found to have values in an interval $d_{\min} \leqslant d_p(\hat{x}) \leqslant d_{\max}$.

(b) *Multifractals* (Mandelbrot 1989)

There is a considerable body of work, both theoretical (for example Halsey *et al.* 1986) and experimental (Jensen *et al.* 1985), which deals with multifractal chaotic attractors. This work either characterizes a fractal measure in terms of a one-parameter family of dimensions $\{d_q | q \in \mathbb{R}\}$ (Renyi 1970; Mandelbrot 1982; Hentschel & Procaccia 1983; Grassberger 1983), where

$$d_q = \frac{1}{q-1} \lim_{\epsilon \to 0} \left[\frac{\ln \left[\int d\mu(\hat{x}) \, \mu[B_\epsilon(\hat{x})]^{q-1} \right]}{\ln \epsilon} \right], \tag{9}$$

or seeks to model it as a collection of singularities such that the set of points where the measure is singular with strength α has Hausdorf dimension $f(\alpha)$ (Halsey *et al.* 1986). This picture leads to a statistical mechanical formalism whereby $f(\alpha)$ is related to d_q via a Legendre transform

$$d_q = \frac{1}{q-1} [q\alpha(q) - f(\alpha)]. \tag{10}$$

If $\mu[B_\epsilon(\hat{x})]$ is independent of $\hat{x}$ almost everywhere with respect to μ the whole family, $\{d_q\}$, collapses to a single value, d_p, the pointwise dimension. More generally, equation (9) defines exponents by averaging local scaling behaviour over the whole attractor. In the particular case of d_2 that is well-known as the correlation dimension (Grassberger & Procaccia 1984)

$$d_2 = \lim_{\epsilon \to 0} \left[\frac{\ln \left[\int d\mu(\hat{x}) \, \mu[B_\epsilon(\hat{x})] \right]}{\ln \epsilon} \right] \tag{11}$$

there is a close analogy with the pointwise dimension. Instead of looking for an exponent that is independent of position, the correlation dimension is an exponent that gives the average scaling properties of $\mu[B_\epsilon(\hat{x})]$. In the following sections of this paper some of the implications of performing this average while at the same time being unable to take the limit $N \to \infty$, or, equivalently $\epsilon \to 0$, will be considered.

6. MAXIMUM LIKELIHOOD ESTIMATES OF $d_p(\hat{x})$ AND d_2

This section recalls the work of Takens (1984) on the use of maximum likelihood methods for the estimation of correlation dimension.

We begin with a simplified version of Taken's algorithm, which leads to an estimate of the local exponent $d_p(\hat{x})$. This has four steps:

(i) a centre, $\hat{x}$, is chosen from the trajectory;
(ii) a random set of N_B points that lie within the ball $B_\epsilon(\hat{x})$ is chosen;

(iii) the radial distances of these points from the centre are scaled to lie in the interval $[0, 1]$

$$r_i = |\hat{x}_i - \hat{x}|/\epsilon, \tag{12}$$

where $|...|$ denotes the euclidean distance;

(iv) the mean, $\langle \ln r \rangle$, is calculated for the distribution of radii.

Takens shows that with the scaling ansatz $\mu[B_\epsilon(\hat{x})] \sim \epsilon^{d_p(\hat{x})}$, the maximum likelihood estimate for the exponent $d_p(\hat{x})$ is given by

$$\langle \ln r \rangle = -\alpha^{-1}. \tag{13}$$

The standard error for this estimate is $(N_B^{\frac{1}{2}}\alpha)^{-1}$. A consequence of this approach is that α^{-1} is an optimum estimate for $d_p(\hat{x})^{-1}$ in the sense that it has minimum variance. In the form of equation (13) it is also an unbiased estimate. Taking the reciprocal introduces a bias that is reflected in our figures by asymmetric error bars (corresponding to plus or minus one standard error in the estimate for $d_p(\hat{x})^{-1}$).

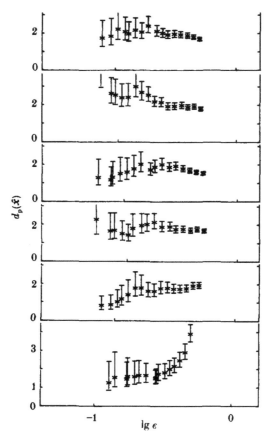

FIGURE 5. Maximum likelihood estimates of local exponents $d_p(\hat{x})$ for the attractor shown in figure 2.

Takens's algorithm for the estimate of d_2 differs from the above in a single detail · when compiling the distribution of scaled radii in step (ii), the centre of the ϵ-ball, $\hat{x}$, is also allowed to vary at random. This provides the additional average required in equation (11).

As an example of this approach we present in figure 5 the results of estimating $d_p(\hat{x})$, for six randomly chosen centres, with the data shown in figures 1 and 2. The results are consistent with an exponent *ca.* 2 for the selection of centres chosen. However, they do not provide an accurate estimate of the non-integer parts of these exponents. Of course, if d_2 were to be calculated it may be possible to obtain larger values of N_B for a given radius, and therefore tighter error bounds. However, the additional averaging procedure may itself introduce systematic errors. This point will be discussed in the next section.

7. LIMITS AND LIMITATIONS

There are two limits which must be considered when calculating the dimension of the natural measure: (a) the number of points, $N \to \infty$ and (b) the radius $\epsilon \to 0$. These two are related since $N \sim \epsilon^{d_0}$ for small ϵ, where $d_0 = \lim_{q \to 0} d_q$.

(a) *The limit* $N \to \infty$

One implication of the limit $N \to \infty$ is that the experiment must continue for infinite time. Imagine that such an experiment is done with the intention of studying the detailed fractal structure of the attractor shown in figure 2. The trajectory is constructed from the time series and as it traces out its path in the embedding space intersections with section 1 in figure 2 are found. The vertical component of each intersection is then calculated because, in this case, it carries most of the fractal information. This is done by forming the weighted sum of the contents of the delay register, see equation (3), with the weights plotted in figure 6. As the experiment progress this weighted sum must take values that distinguish the detailed structure of the fractal on arbitrarily fine scales. We conclude that fractal information in a time series such as the one in figure 1 is carried in the structure of coherent patterns that occur over arbitrarily long time scales. To detect such long time coherence it is important to control the experimental conditions over the whole length of the experiment. If this is not

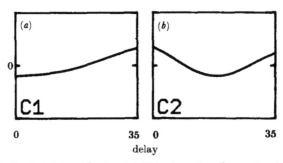

FIGURE 6. Vertical (a) and horizontal (b) basis vectors for section 1 of figure 2

done then increasing N to better approximate the limit may actually be counter productive.

Recently, Smith (1989) has found that, for reliable estimates of the correlation dimension, N must at least satisfy

$$N \geqslant 42^M, \tag{14}$$

where M is the greatest integer less than the dimension of the set. This result may be used as a self-consistent criterion to judge the suitability of the length of a data set for estimating d_2. In §8, however, we describe a method that gives an independent estimate of M. Equation (14) also provides the timescale, T_{exp}, over which the experimental conditions must be controlled. If we assume that T, the dominant period of the system, is the mean time between intersections with a chosen surface, then

$$T_{exp} \geqslant 42^{M-1}T, \tag{15}$$

which shows that the problem of control becomes acute as the dimension of the attractor increases.

(b) *The limit* $\epsilon \to 0$

Ignoring any lack of data, there are two factors that affect the degree to which the limit $\epsilon \to 0$ can be approximated. These are the finite precision of the measurement and external noise. Although different in origin their affects are similar.

For example, consider the calculation of the vertical components of section 1. For the weighted sum over the contents of the delay register to resolve the fine structure of the fractal, the individual measurements in the register must have sufficiently high precision. We note, however, that it is not uncommon to find data collected with a 9-bit A/D converter giving a precision of *ca*. 3 significant figures. A 16-bit A/D converter, currently state-of-the-art, gives *ca*. 5 significant figures.

The singular value technique described in §3 can help in two ways. Firstly, some of the noise can be projected onto a subspace orthogonal to the one containing the trajectory, and secondly, the singular values in the noise floor define the length scale in the embedding space at which noise processes and precision become important.

There are 'macroscopic' effects, not related to the fractal scales that arise because the limit $\epsilon \to 0$ cannot be reached. These are best seen by considering the scaling properties of the measure $\mu[B_\epsilon(\hat{x})]$ as a function of radius. One effect, noted by Guckenheimer (1984), involves the choice of a centre, $\hat{x}$, near a boundary of the attractor. Consider the example of a uniform distribution on a large square with a centre $\hat{x}$ chosen to be ϵ_0 from the edge of the square. For $\epsilon < \epsilon_0$, the measure $\mu[B_\epsilon(\hat{x})]$ is equal to the area of a circle of radius ϵ, $\pi\epsilon^2$. As ϵ increases through ϵ_0 the measure becomes the area of a segment of a circle. For large ϵ, as $\epsilon_0/\epsilon \to 0$, the area of this segment approaches $\frac{1}{2}\pi\epsilon^2$. Therefore, for $\epsilon > \epsilon_0, \mu[B_\epsilon(\hat{x})]$ does not scale simply as ϵ^2, but has an ϵ-dependent prefactor.

The situation becomes worse if the attractor is thin; in this case the exponent of the scaling can change. For example, consider a uniform measure on a long thin rectangle of width $2\epsilon_0$. Putting $\hat{x}$ at the centre of this rectangle we find that $\mu[B_\epsilon(\hat{x})]$ $\sim \pi\epsilon^2$ when $\epsilon < \epsilon_0$. However, for large ϵ, as $\epsilon/\epsilon_0 \to 0$, the rectangle takes on the appearance of a pencil, the measure will then scale linearly with ϵ because the dominant contribution to the scaling comes from the length of the pencil that lies inside the ball.

Another effect involves the metric used in the embedding space. The trajectory, and therefore the attractor, is assumed to exist in a nonlinear manifold. This can twist in such a way that points that are well separated when using a metric appropriate to the manifold appear to be close in the euclidean metric of the embedding space. For example, consider an elliptic limit cycle whose eccentricity is close to unity. In this case points that are maximally separated when measured along the elliptic limit cycle may be very close in the plane if they lie near opposite ends of the minor axis. Presently, there is little choice but to use the euclidean metric of the embedding space as there is generally insufficient information about the manifold for a more natural metric to be constructed. This constraint makes it possible that distinct pieces of the attractor can be found within the same ball, with profound effects on the scaling of $\mu[B_\epsilon(\hat{x})]$.

The occurrence of such 'macroscopic' effects has serious implications when the local scaling behaviour is averaged over the whole attractor as in the definition of the correlation dimension. Enlarging ϵ to improve the statistics by increasing the number of points in the ball will ultimately enhance the problem of averaging spurious exponents and ϵ-dependent prefactors. It is not clear that these effects will in any way 'average out' to produce a meaningful result.

8. LOCAL ANALYSIS

In §7 we showed that 'macroscopic' effects can create anomalies in estimates of the local exponents $d_p(\hat{x})$. These are difficult to interpret because the calculation of $\mu[B_\epsilon(\hat{x})]$ requires a simple isotropic average that is insensitive to the local geometry. More detail can be obtained from the local anisotropy of the measure, which may be studied with a modification of the singular value analysis (Broomhead *et al.* 1987*a*, *b*). Formally, we use the local covariance matrix Ξ_ϵ for the ball $B_\epsilon(x)$:

$$\Xi_\epsilon = \frac{1}{\mu[B_\epsilon(x)]} \int_{B_\epsilon(x)} d\mu(x') (x' - x) (x' - x)^{\mathrm{T}}. \tag{16}$$

Information about the anisotropy of the measure when restricted to the ball is contained in the eigenvectors and spectrum of this matrix:

$$\Xi_\epsilon b_i = \sigma_i^2(\epsilon) b_i. \tag{17}$$

This analysis is analogous to the global method described in §3 (compare equation (17) with equation (1)), where the eigenvectors can be thought of as the axes of inertia of a mass distribution within the ball, and the local singular values $\{\sigma_i\}$ give the width of the mass distribution projected onto these axes. Three obvious differences from the global method are (i) the vectors are taken relative to the centre of the ball rather than $\langle v \rangle \delta$, (ii) the average is taken over only those points within $B_\epsilon(x)$ and (iii), there are now additional parameters which may be varied; the position and radius of the ball.

In practice, when analysing time series data, a centre, x, is chosen as a point on the trajectory and a local matrix with rows $(x_i - x)^T$ is constructed by selecting points, x_i, on the trajectory such that $|x_i - x| < \epsilon$. The same selection procedure is done when estimating $\mu[B_\epsilon(\hat{x})]$ for dimension calculations, but in addition to counting the number of points within $B_\epsilon(x)$, the points themselves are retained. The local singular vectors and spectrum are obtained numerically by singular value decomposition of the local matrix. In fact, to avoid accumulating error with unnecessary layers of computation we perform the SVD of the local matrix by using the coordinates of the delay register, i.e. with rows $(\hat{x}_i - \hat{x})^T \to (x_i - x)^T$.

The local structure of a chaotic attractor generally consists of directions in which the invariant measure is smooth (corresponding to the expanding unstable direction in figures 2 and 3) and directions in which the measure is fractal (corresponding to the contracting stable direction in figures 2 and 3).

The smooth directions may be thought of as an unstable manifold. For small ϵ, the difference vectors $(x_i - x)^T$ approximate vectors in the tangent plane of this manifold centred at x. As ϵ increases, the manifold finds room to curve within $B_\epsilon(x)$, and components of $(x_i - x)$ will develop that are normal to the tangent plane. This behaviour can be identified by observing the scaling properties of the local singular spectra as the radius of the ball is increased. It can be shown that as $\epsilon \to 0$, a part of the spectrum will consist of a set of singular values that scale linearly with ϵ. This set corresponds to singular vectors that span the tangent space. If the manifold has a quadratic tangency with the tangent space then there will be an additional set of singular values which scale as ϵ^2 (the exponent will generally be the order of the tangency). The singular vectors corresponding to these are normals to the tangent space indicating the directions in which the manifold curves.

In the fractal directions the singular values will not be smooth. In the simplest case of one singular value, σ, corresponding to a Cantor set in one direction the function $\sigma(\epsilon)$ is a 'Devil's staircase', that is, a non-decreasing function that is nowhere differentiable. For attractors with more than one fractal direction the interaction between the corresponding singular values should produce more complicated forms.

In §2 it was remarked that the fractal direction of an attractor is often very compressed. A consequence of this is that very long time series are required to provide enough points inside a ball, sufficiently small to resolve the fractal structure. This situation is rather like the example of the long thin rectangle given in §7. In the present case when the radius of the ball is larger than the width of the fractal the corresponding singular value is essentially independent of ϵ. It must then be established whether or not the constant value is significantly greater than the noise floor of the singular spectrum, which is also independent of ϵ.

We define the local 'geometric dimension' (referred to as the 'topological dimension' in Broomhead et al. 1987 a, b) as the rank of the local matrix as $\epsilon \to 0$. This consists of the sum of the number of singular values that scale as ϵ^1 and the number of singular values that correspond to fractal directions. In practice there may be a number of constant singular values, corresponding to fractal directions and/or directions in which the attractor happens to be narrow, these can be used in the estimate of the geometric dimension. Geometric dimension uses the intuitive idea of dimension as the 'number of independent directions' in the space; because of this it is less demanding on the quantity of data than definitions that rely on finding a scaling exponent. It can be used as an estimate of m, the dimension of the manifold subjected to the embedding process described in §4. It is also related to Smith's M used in equation (14) and equation (15).

The local basis generated by this method can be used to study local dynamics near a fixed point of the experimental flow (Broomhead & King 1986*b*). A collection of such bases can provide a natural, global representation of the manifold as a set of charts to the corresponding tangent spaces. We have not attempted to do this with experimental data. However, the same technique can be applied to the study of solutions confined to finite-dimensional invariant manifolds of partial differential equations (Broomhead *et al.* 1989).

(a) *An example*

The data used for this example comes from a study of convection in liquid gallium subject to a horizontal temperature gradient and a horizontal, transverse magnetic field. It consists of a time series of voltages, v, taken from a thermocouple probe. Figure 7 shows the reconstructed trajectory projected onto the first two global singular vectors together with the indicated cross section. The system appears to be in locked state with a period *ca.* 10 times the dominant period of the signal. It is possible, however, from the evidence in the section, that the orbit lies on a torus rather than a limit cycle.

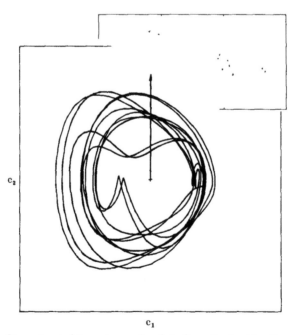

c_2

c_1

FIGURE 7. Reconstructed trajectory using data from the liquid gallium convection experiment with cross section taken at the place indicated

Figure 8 shows a plot of the maximum likelihood estimate of the local exponent, $d_p(\hat{x})$, as a function of ϵ. This particular choice of centre, which was made at random, provides an illustration of the macroscopic effects on scaling calculations discussed in §7. In the plot there are three distinct regions: (A) $\lg \epsilon < -0.85$ where, accepting that the statistical accuracy is low, the exponent is *ca.* 1; (B) $-0.85 < \lg \epsilon < -0.725$ where the exponent increases rapidly to a value of *ca.* 2 and (C) $\lg \epsilon > -0.725$ where the exponent seems to saturate at *ca.* 1.5 with good statistical accuracy.

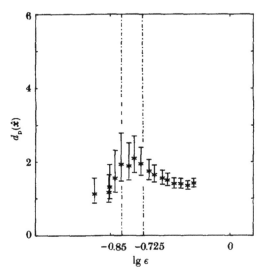

FIGURE 8. Maximum likelihood estimates of a local exponent $d_p(\hat{x})$ for the attractor shown in figure 7.

For comparison, figure 9 shows a corresponding plot of the local singular spectra as a function of ϵ. The three regions are again apparent. In region A there is one singular value scaling as ϵ^1 together with one constant singular value above the noise floor. The noise floor is indicated by the dash–dot line in figure 9 and is taken from the global analysis. It corresponds to *ca.* 9 bits of resolution of the A/D converter. This noise floor is consistent with the dominant contribution being quantization noise. We interpret the behaviour of the singular values in region A as meaning that the ball at these radii contains one 'strand' of the limit cycle or torus. This agrees with $d_p(\hat{x}) \approx 1$, because we have essentially the 'pencil' example given in §7, but suggests that the torus hypothesis might be worthy of further investigation. If the attractor is a narrow torus we would expect to see two constant singular values. The fact that we do not suggests a flat structure, possibly a flattened torus or alternatively a limit cycle distorted by a slight drift in the control parameters for the experiment.

In region B of figure 9 an additional singular value appears to be growing linearly and this saturates in region C. This behaviour is consistent with the enlarged ball including, from nearby in the embedding space, an extra strand of

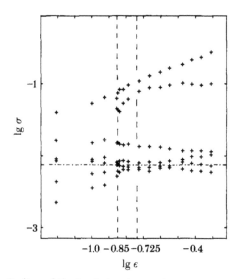

FIGURE 9. Scaling of the local singular spectrum corresponding to figure 8

the limit cycle. In region B, as soon as the ball is large enough to include a point from the second strand a corresponding singular value, which measures the width of the distribution, changes its value discontinuously. It then increases linearly until the full width of the new strand is included. Beyond this point the singular value saturates. It is possible, by using figure 9, to estimate the width of the two strands included in the ball. The first strand has a width $\Delta_1 \approx 0.03$ standard deviations of v, estimated by extrapolating to the value of ϵ at which the constant and linearly growing singular values are equal. The width of the second strand $\Delta_2 \approx 0.05$ standard deviations, is obtained from the width of region B.

9. CONCLUSION

In this paper a set of techniques has been described that, when applied to time-series data, give geometric information about the dynamical system from which the data was obtained. These ideas have been discussed in the context of the static geometry of fractal measures on invariant sets. They can also be used to obtain dynamical information such as characteristic exponents and the eigenvalues associated with fixed points of the dynamical system. It is possible to define a fractal dimension of an attractor in terms of its characteristic exponents (Eckmann & Ruelle 1985). Recent work suggests that, in a noisy environment, this dimension equals the pointwise dimension when it exists (Ladrappier & Young 1987).

These methods provide information in a form that relates directly to the mathematics of nonlinear systems. They can be used therefore, to check the relevance of theorems to the physics of the experimental system. Thus there are theorems that relate the existence of various fractal invariant sets to the ratio of eigenvalues at saddle points with homoclinic orbits (Shilnikov 1965; Glendinning & Sparrow 1984; Gaspard *et al.* 1984). Our preliminary results suggest that this approach is more robust to experimental imperfections than the fractal dimension calculations we have discussed.

We thank Don Hurle and Kathy McKell for kindly providing data from their liquid gallium convection experiment and Greg King for all his contributions to the work.

REFERENCES

Argoul, F., Arneodo, A. & Richetti, R. 1987 *Phys. Lett.* A **120**, 269.

Arneodo, A. & Holschreider, M. 1987 Fractal dimensions and homeomorphic conjugacies. (Preprint.)

Bergé, P., Pomeau, Y. & Vidal, C. 1984 *Order within chaos.* New York: John Wiley.

Broomhead, D. S. & King, G. P. 1986a *Physica* **20**D, 217.

Broomhead, D. S. & King, G. P. 1986b In *Nonlinear phenomena and chaos* (ed. S. Sarkar). Bristol: Adam Hilger.

Broomhead, D. S., Jones, R. & King, G. P. 1987a *J. Phys.* A **20**, L563–L569.

Broomhead, D. S., Jones, R., King, G. P. & Pike, E. R. 1987b In *Chaos, noise and fractals* (ed. E. R. Pike & L. A. Lugiato). Bristol: Adam Hilger.

Broomhead, D. S., Newell, A. C. & Rand, D. A. 1989 (In preparation.)

Constantin, P., Foias, C. & Teman, R. 1985 *Memoirs of the AMS* **53**, no. 413.

Devijver, P. & Kittler, J. 1982 *Pattern recognition: a statistical approach.* London: Prentice-Hall.

Eckmann, J.-P. & Ruelle, D. 1985 *Rev. Mod. Phys.* **57**, 617–656.

Farmer, J. D., Ott, E. & Yorke, J. A. 1983 *Physica* **7**D, 153.

Fischer, P. & Smith, W. R. (ed.) 1985 *Chaos, fractals and dynamics.* New York: Marcel Dekker Inc.

Gaspard, P. & Nicolis, G. 1983 *J. statist. Phys.* **31**, 499.

Gaspard, P., Kapral, R. & Nicolis, G. 1984 *J. statist. Phys.* **35**, 697.

Glendinning, P. & Sparrow, C. 1984 *J. statist. Phys.* **35**, 645.

Golub, G. H. & van Loan, C. F. 1985 *Matrix computations.* Baltimore: Johns Hopkins University Press.

Grassberger, P. 1983 *Phys. Lett.* **97**A, 227.

Grassberger, P. & Procaccia, I. 1984 *Physica* **13**D, 34.

Guckenheimer, J. & Holmes, P. 1983 *Nonlinear oscillators, dynamical systems and bifurcations of vector fields.* New York, Berlin: Springer-Verlag.

Guckenheimer, J. 1984 *Contemp. Math.* **28** 357.

Halsey, T. C., Jensen, M. H., Kadanoff, L. P., Procaccia, I. & Sharaiman, B. I. 1986 *Phys Rev.* A **33**, 1141.

Hentschel, H. G. E. & Procaccia, I. 1983 *Physica* **8**D, 435.

Hirsch, M. W. 1976 *Differential topology.* New York, Berlin: Springer-Verlag.

Jensen, M. H., Kadanoff, L. P., Libchaber, A., Procaccia, I. & Stavans, J. 1985 *Phys. Rev. Lett.* **55**, 2798.

King, G. P., Jones, R. & Broomhead, D. S. 1987 *Nucl. Phys.* B (Suppl.) **2**, 379–390.

Ledrappier, F. & Young, L.-S. 1987 Dimension formula for random transformations. (Preprint.)

Mandelbrot, B. B. 1982 *The fractal geometry of nature.* San Francisco: W. H. Freeman & Co.

Mandelbrot, B. B. 1989 In *Fluctuations and pattern formation* (Cargèse 1988) (ed. H. E. Stanley & N. Ostrowsky). Dordrect, Boston: Kluwer. (In the press.)

Meyer-Kress, G. (ed.) 1986 *Dimensions and entropies in chaotic systems.* New York, Berlin: Springer-Verlag.

Packard, N. H., Crutchfield, J. P., Farmer, J. D. & Shaw, R. S. 1980 *Phys. Rev. Lett.* **45**, 712.

Peitgen, H.-O. & Richter, P. H. 1986 *The beauty of fractals.* New York, Berlin: Springer-Verlag.

Pike, E. R., McWhirter, J. G., Bertero, M. & de Mol, C. 1984 *IEE Proc.* F **131**, 660.

Poincaré, H. 1899 *Les methodes nouvelles de la mechanique celeste.* (3 volumes.) Paris: Gauthier-Villar

Renyi, A. 1970 *Probability theory.* Amsterdam · North Holland.

Rössler, O. E. 1977 *Bull. math. Biol.* **39**, 275.

Roux, J. C., Simoyi, R. H. & Swinney, H. L. 1983 *Physica* **8D**, 257–266.

Shilnikov, L. P. 1965 *Soviet Math. Dokl.* **6**, 163–166.

Smith, L. A. 1989 *Phys. Lett.* A (In the press)

Takens, F. 1981 In *Lecture notes in mathematics*, vol. 898 (ed. D A. Rand & L.-S. Young), p. 366. New York, Berlin. Springer-Verlag.

Takens, F. 1984 In *Lecture notes in mathematics*, vol 1125, p. 99. New York, Berlin: Springer-Verlag.

Thompson, J. M T 1989 *Proc. R. Soc. Lond.* A **421**, 195–225.

Tomita, K. & Kai, T. 1978 *Suppl. Prog. theor. Phys.* **64**, 280–294

Young, L.-S 1982 *Ergod. Theory Dynam. Syst.* **2**, 109.

Diffusion-controlled growth

By R. C. Ball, M. J. Blunt and O. Rath Spivack

Cavendish Laboratory, Madingley Road, Cambridge CB3 0HE, U.K.

The conditions of diffusion-controlled growth are outlined and the observed importance of anisotropy is discussed through a tentative flow diagram.

A crucial role is played by the forwardmost tips, which lead to growth. The nature of the singularity in their growth rate determines the overall fractal dimension. This has been estimated in two dimensions from effective cone-angle models, which work well for the most extreme anisotropic growth and can be augmented into a self-consistent approximation for the isotropic fractal case.

The way in which the tip growth rate singularity is limited by finite tip radius is also a key ingredient. For diffusion-limited solidification where it is set by competition with surface tension, this significantly changes the form of the equivalent model with a fixed (e.g. lattice spacing) imposed tip scale.

The full distribution of growth rates everywhere provides a much richer problem. We show new data and examine the consistency of how sites can evolve from the regions of high growth rate where they are born, into well-screened regions devoid of further growth.

1. Introduction

The key element relating different analogues of diffusion-controlled growth (or aggregation) is that the local growth rate is determined by the flux of a field that obeys Laplace's equation, with the growth surface presenting an equipotential boundary condition. This is evidently realized by true material diffusion in the limit that the growth advances slowly enough that the concentration field ahead of it is quasi-static, and the extreme case of this is represented by the original diffusion limited aggregation (DLA) computer model of Witten & Sander (1981) where only one particle approaches the aggregate at any one time.

It is valuable to distinguish between stochastic models, in which the interface advances locally in non-zero steps with a *probability* given by the local flux, and causal models in which the interface advances continuously at a *speed* given by the flux. Various techniques have been found to systematically reduce the stochastic element of the discrete stepping models (Tang 1985; Kertesz & Vicsek 1986; Nittman & Stanley 1986), so that local noise in the growth rate is a parameter open to systematic study.

The causal continuum version exhibits the Mullins–Sekerka instability. A small-amplitude surface corrugation of wavevector k grows exponentially (at rate k) with the mean advance of the interface (Mullins & Sekerka 1963 a, b), and infinitely sharp cusps can develop after only a finite advance (Shraiman & Bensimon 1984). Hence there must be a change in the physics at short length scales, giving an effective spatial cut-off to the domain of diffusion control. Lattice simulations incorporate this directly with a fixed cut-off, but we will return to this issue in §5.

2. ANISOTROPY

It is now very clear that for large enough growths, underlying anisotropy can be crucial for the large-scale morphology. It is mostly *rigid* anisotropy that has been considered, in the sense that we are talking about discrete favoured directions such as lattice axes that cannot meander continuously as the cluster grows. This occurs gratuitously in lattice-based computer simulations, but is also a feature of real dendritic solidification where the local crystalline order can be faithfully transmitted through several stages of ramification (Grier *et al.* 1986).

Figure 1 gives a tentative pictorial symmary of various peoples' discoveries in two dimensions. For the square lattice ($m = 4$) the instability of the isotropic fractal with respect to anisotropy was first observed by Ball & Brady (1984), the clearest dendrites have been grown by Meakin (1987 a) and the behaviour in the needle-growth régime for this and other anisotropies has recently been developed by Meakin & Eckmann (1989). The meaning of my disorder axis is less than clear, but draws inspiration from the 'noise reduction' methods first introduced by

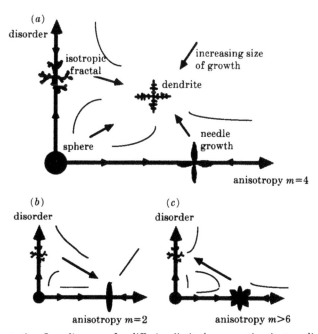

FIGURE 1. Tentative flow diagrams for diffusion-limited aggregation in two dimensions. The arrows indicate the direction of evolution as the clusters grow large. (a) In fourfold ($m = 4$) bias (e.g. square lattice) there are four fixed points of which only the dendrite is stable with respect to both anisotropy and noise. (b) In two-fold bias ($m = 2$) the dendrite does not side branch and becomes a needle. (c) For high enough m, anisotropy cannot sustain its tips and the dendrite becomes isotropic ($m > 6$).

Kertesc & Vicsek (1986), which they found to dramatically highlight the importance of anisotropy. The case of $m = 2$ was studied by Ball *et al.* (1985) to verify the cone-angle approximation (below) and always showed evolution towards the compact needle morphology.

What happens with high-order anisotropy is somewhat more controversial. Calculations (Ball 1986) indicate that no more than six fingers or needles can simultaneously be stable, and simulations by Meakin (1987a, b) with high noise reduction appear to indicate that the isotropic fractal behaviour is approached rather than stabilization of needles in a spontaneously selected subset of the possible directions. Neither study is, however, definitive.

3. SIGNIFICANCE OF GROWTH TIPS

The tips of a diffusion-limited growth are crucial and the probability p_{tip} of growth occurring there determines the mass–radius relation for the whole growth. For a DLA-type model where the particles are of size a the average rate of increase of the extremal radius R_{ext} of the cluster is simply given by (Ball & Witten 1984)

$$\langle \mathrm{d}R_{\text{ext}}/\mathrm{d}N \rangle \approx a p_{\text{tip}}. \tag{1}$$

If we anticipate a power-law singularity of the growth probability at the tip so that

$$p_{\text{tip}} \approx (a/R_{\text{ext}}) \alpha_{\text{tip}}, \tag{2}$$

then by comparison with the fractal scaling law $N \sim R^{D}$ we find immediately that

$$D = 1 + \alpha_{\text{tip}}. \tag{3}$$

Thus α_{tip} determines the fractal dimension. It has also been commonly assumed that the tips correspond to the most active sites so that $\alpha_{\text{tip}} = \alpha_{\text{min}}$.

This leading behaviour of the tips may shed light on why anisotropy can be so important. For example on the square lattice a tip advancing along a diagonal direction requires $\sqrt{2}$ times as many growth steps as one advancing the same distance along an axial direction. This assumes that DLA tips follow straightest possible paths, which by observation is not correct: however, their paths are statistically straight in the sense that simulation measurements (Meakin *et al.* 1984) have shown that on average the path length connecting two points *along* the cluster is directly proportional to their euclidean separation.

4. ESTIMATES OF α_{tip} IN TWO DIMENSIONS

An important feature of diffusion-limited growth is the strong screening of the interior from the incoming field, which leads naturally to the suggestion that only the outer envelope of the growth and not the detailed structure inside the fiords is relevant in determining α_{tip}. In two dimensions, the only simple shape of absorber that is self-similar about its tip and gives a pure power-law singularity in the growth rate there, is a cone or wedge (Turkevich & Scher 1985; Ball *et al.* 1985). In terms of the cone angle θ (see figure 2) one has simply

$$\alpha_{\text{tip}} = \pi/2\theta. \tag{4}$$

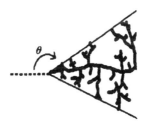

FIGURE 2. Geometry of the cone-angle model.

For isotropic DLA with $D = 1.71$ the above relation implies an equivalent effective angle $\theta_{\text{eff}} \approx 125°$.

To obtain a test case where angles could unambiguously be defined from cluster geometry, Ball *et al.* (1985) studied square lattice DLA with sticking probabilities biased in favour of growth along one axis. Asymptotically the growths developed into needles whose advance in both the long and short directions was consistent with the exponents given by equation (4) with the *observed* limiting angles $\theta = \pi$ and $\theta = \frac{1}{2}\pi$ respectively.

The cone angle estimate of α_{tip} has also been developed into a complete (but approximate) theory of DLA in two dimensions (Ball 1986). The first ingredient is a relation between α_{tip} and the global morphology of the cluster in terms of the number of principal growth tips n, by requiring that the sides of the (equal angle) wedges close to form a polygon:

$$n(2\theta - \pi) = 2\pi. \tag{5}$$

The second ingredient is the stability of such a structure under growth, with respect to some of the principal tips falling behind others and being screened out: conformal mapping arguments gave

$$(\tfrac{1}{2}n - 1)\,\alpha_{\text{tip}} = 1 \tag{6}$$

for the maximum n that could be sustained. The selected value $n = 2 + \sqrt{8}$ represents an analytic continuation of unclear status, but the value for the fractal dimension $D = 1 + \frac{1}{2}\sqrt{2}$ is rather impressive.

For the fractal dimension of DLA in two dimensions the mean field result $D = d - 1$ (Ball *et al.* 1984) reduces to $D = 1$. The best rigorous bound is $D \geqslant \frac{3}{2}$ (by using $\theta_{\text{eff}} \leqslant \pi$) and a popular formula (Muthukumar 1983) gives $D = \frac{5}{3} \approx 1.667$. The cone-angle and tip stability value discussed above gives numerically $D \approx 1.707$, now three standard deviations from the most precise simulation estimate (Meakin 1989) $D = 1.713 \pm 0.004$.

5. RELATION TO SOLIDIFICATION

This is the one section of this contribution where we will consider diffusion-limited growth with a cut-off that is not constant, but depends on the local growth rate. Real solidification can be limited by the diffusion of either latent heat or compositional excess (created by solidification at a composition different from the mother liquid) and the crucial feature cutting off the Mullins–Sekerka instability is surface tension (see Langer 1980). The Laplace field at the surface of the growth is set to a value not quite equipotential, but that has a capillary offset proportional to the local surface curvature, and equal to it in appropriately scaled units (Ball 1989). Under these conditions the highest unstable wavevector of surface corrugation on a nearly planar surface is given by $k_c = \sqrt{v}$, where v is the local growth velocity without corrugation. This leads to the famous selection criterion (see, for example, Langer 1980; Huang & Glicksman 1981) that tip radius a and velocity are related through

$$va^2 = \text{const.} \tag{7}$$

There has been continuing and deep debate over the precise value of the constant and its crucial sensitivity to anisotropy for the case of a nearly parabolic needle. The fact that stable needle solutions cannot be found in the absence of anisotropy, a surprise in solidification, is less surprising if the phase diagram of figure 1 for DLA is correct. However, for the purpose of the present argument we will only need the existence of a constant in equation (7), not its value.

The strong dependence of tip radius on growth velocity can totally change the nature of the growth. Here we show that the most closely related lattice model with a *fixed* cut-off is one in which the relation between growth and local flux (at the level of a fixed cut-off) is nonlinear.

We begin by assuming that the singularity in the growth rate at a leading tip is distributed, in the sense that the total diffusing flux being absorbed within a region of radius $r > a$ of the actual tip obeys

$$\mu(r) = \mu(c)\,(r/c)^{\alpha_{\text{tip}}}, \tag{8}$$

where c is our *fixed* reference scale.

We now match the velocity from the selection criterion (7) to the advance rate dictated by the flux density onto the actual tip of radius a:

$$v = a^{-2} = \mu(a)/a^{d-1} = \mu(c)\,c^{-\alpha_{\text{tip}}}\,a^{\alpha_{\text{tip}}+1-d}, \tag{9}$$

from which we can eliminate the tip radius a and find

$$v = (\mu(c)\,c^{-\alpha_{\text{tip}}})^\eta, \tag{10}$$

where the exponent is given by

$$\eta = 2/\alpha_{\text{tip}} + 3 - d. \tag{11}$$

The latter two equations constitute the nonlinear growth rule of the generalized dielectric breakdown model (DBM) introduced by Niemeyer *et al.* (1984). They regarded η as an arbitrary model parameter, and showed by computer simulation that fractal clusters were generated whose scaling properties depended continuously on η. Here η must be self-consistently selected so that condition (11) is obeyed, by using the value of α_{tip} appropriate to the corresponding DBM growths.

We have seen in this section how non-trivial physics for the cut-off can change the basic form of the equivalent fixed cut-off simulation, reinforcing the crucial nature of the growth tips. The remainder of this paper, however, will return to the case where we have a linear relation between growth rate and flux at a fixed cut-off scale, as in the DLA model.

6. THE DISTRIBUTION OF GROWTH

We now turn our attention to the complete distribution of growth rates (or probabilities) over the whole of a DLA or related cluster (Halsey *et al.* 1986). In the limit of a large cluster of linear dimension R viewed on the scale of much smaller regions of linear dimension b this is now well known to have multifractal form (Mandelbrot 1982)

$$(R/b)^{f(\alpha)} \text{ regions have total flux } \mu = (R/b)^{-\alpha}. \tag{12}$$

For convenience in what follows we will choose the convention that $b = 1$ corresponds to individual sites, and take the fluxes to have been normalized so that their total sum is unity. Our latest numerical results for $f(\alpha)$ in two-dimensional DLA are plotted in figure 3.

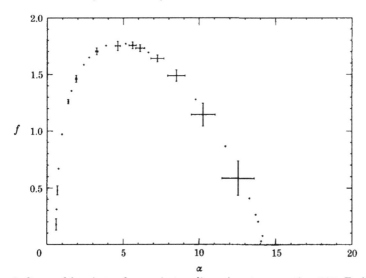

FIGURE 3. Curve of f against α for DLA in two dimensions (see equation (12)). Each data point was obtained numerically by specifying the value of $q = \mathrm{d}f/\mathrm{d}\alpha$ and computing f and α independently as generalized entropies (see Ball & Rath Spivack 1989). A scaling superposition of data from clusters of 10000–100000 particles was used; the clusters were grown on square lattice but with partial second neighbour sticking that significantly suppresses the influence of lattice anisotropy.

The total number of sites is simply given by

$$N = \int \mathrm{d}\alpha \, R^{f(\alpha)}, \tag{13}$$

which in the limit of large enough R is dominated by its steepest descent point at

$$q(\alpha) = \mathrm{d}f/\mathrm{d}\alpha = 0, \tag{14}$$

whereas the total rate of growth is given by summing all the fluxes

$$dN/dN = \int d\alpha \, R^{f(\alpha)-\alpha} \quad (= 1) \tag{15}$$

and is dominated by $q(\alpha) = 1$ (where it must happen that $f = \alpha$ to match unity).

Growth corresponds to the creation of new sites and the flux onto these will be of the same order of magnitude, hence the value of α will be the same, as that onto the sites from which they grew. It follows that most new sites are born with $\alpha = \alpha_1$ whereas the majority of all sites have $\alpha = \alpha_0$. (The subscripts here refer to the corresponding values of q.) This presents us with the problem of the dynamics of screening: how the growth probability at a site decreases as the cluster evolves, or how sites flow from lower to higher values of α.

7. THE DYNAMICS OF SCREENING

The constraint that sites are conserved gives us useful information. Thinking of N as a 'time' and α as a 'position' variable the equation of conservation gives

$$\partial/\partial N \, R^{f(\alpha)} + \partial/\partial\alpha \, u_T \, R^{f(\alpha)} = R^{f(\alpha)-\alpha}, \tag{16}$$

where u_T is the rate of change $d\alpha/dN$ for a fixed site, averaged over all the sites with given α. Integrating the equation of conservation with respect to α leads immediately to the form

$$\left. \begin{aligned} u_T &= R^{-h_T(\alpha)}, \\ h_T(\alpha) &= \alpha & \alpha \leqslant \alpha_1 \\ &= f(\alpha) & \alpha_1 \leqslant \alpha \leqslant \alpha_0 \\ &= D & \alpha_0 \leqslant \alpha. \end{aligned} \right\} \tag{17}$$

Now we attempt to compare the average evolution above with a direct calculation of the screening of a single site. As pointed out by Halsey (1987) the rate at which a site (labelled by A) screens can be written as

$$d\mu_A/dN = -\sum_B \mu_B^2 \mu_{AB}, \tag{18}$$

where the summation is over all other sites B and μ_{AB} is, to within a numerical factor, the probability that a random walker in the vicinity of B would next touch the cluster at A.

At the level of correct scaling (see Ball & Blunt 1989a) we can reduce the above to

$$u_L = d\alpha/dN = \int d(\ln b) \, \mu(b) \, b^{d-2-\tau_3}, \tag{19}$$

where $\mu(b)$ is the probability of growth within a distance of order b of the point of interest and τ_3 is the minimum of $3\alpha - f(\alpha)$. We define a *typical* site to be one such that $\mu(b)$ scales (as expected) like b^α. For such a typical site one then finds the form $u_L = R^{-h_L(\alpha)}$,

$$\left. \begin{aligned} h_L(\alpha) &= \alpha & \alpha \leqslant \tau_3 + 2 - d, \\ &= \tau_3 + 2 - d, & \alpha \geqslant \tau_3 + 2 - d. \end{aligned} \right\} \tag{20}$$

Identifying h_L and h_T for large α leads to the relation

$$D = \tau_3 + 2 - d, \tag{21}$$

which was first found by Halsey from closely related considerations and is numerically confirmed (Halsey 1987). It is easy to show that this crossover value of α for $h_L(\alpha)$ lies between α_1 and α_0 so that it is only between these values that we have a discrepancy to resolve between u_L and u_T.

Some important but quite atypical sites turn out to have high enough u_L to dominate over the typical ones in the average given by u_T. To handle these sites we have to make a generalization of the distribution (12) implied by $f(\alpha)$, based on an assumption of strong self-similarity. Consider a *site* such that the total flux into a region of scale b around it is given for all $b < R$ by

$$\mu(b) = \exp\left[-\int_{\ln b}^{\ln R} ds\, \alpha(s)\right], \tag{22}$$

where s runs over lengthscales in logarithmic terms. The significance of $\alpha(s)$ is that it is a local effective value of α, in the sense that for b' just slightly less than b we have $\mu(b') = (b'/b)^{\alpha(\ln b)}\mu(b)$. Provided we do not claim this to apply at too fine a resolution of scales, the distribution of values of $\alpha(s)$ should match that of the global values of α, and correlations between different scales (about the same site) should be unimportant if sufficiently strict self-similarity applies. Under these conditions the number of such sites will be given by

$$\exp\int ds\, f(\alpha(s)), \tag{23}$$

where the integration is over all scales within the cluster.

Compared to the definitions of the multifractal distribution (12) with $b = 1$ one sees that our generalized site has μ characterized by an overall value of α given by

$$\alpha_{av} = \langle \alpha(s) \rangle_s \tag{24}$$

and that the total number of such sites is characterized by an overall value of f given by

$$f_{av} = \langle f(\alpha(s)) \rangle_s. \tag{25}$$

The convexity of $f(\alpha)$ is such that f_{av} is always maximized for fixed α_{av} by taking $\alpha(s)$ independent of s, consistent with our definitions for a 'typical' site further above.

Using the above ideas we find that the total flow of sites through α-space is dominated in the region $\alpha_1 < \alpha < \alpha_0$ by sites such that $\alpha(s) = \alpha_1$ for large scales $s > \ln r$ and $\alpha(s) = \alpha_0$ for small scales $s < \ln r$. It is the value of the crossover scale r that determines the corresponding value of $\alpha = \alpha_{av}$, and the rate of change of α for such sites given by equation (19) is dominated by contributions from the scale $b = r$. That these sites maximize the flow through α-space can be understood qualitatively: having $\alpha(s) = \alpha_1$ on outer scales maximizes the rate at which growth (leading to screening) occurs within distance r, whereas having $\alpha(s) = \alpha_0$ on the inner scales maximizes the number of such sites contributing to the average u_T.

The surprising result is that the typical new site, born with $\alpha = \alpha_1$, develops an

$\alpha(s)$ with a crossover to α_0 and it is the crossover-scale that evolves gradually up to the size of the cluster. Only in the sense of an average over scales is there any passage through intermediate values of α. This particular evolution came to light because it dominates the total flow of sites in a certain region of α; however, continuing theoretical investigations (Ball & Blunt 1989b) suggest that the picture is more general, namely that all sites born at $\alpha < D$ exhibit crossover evolution to a corresponding $\alpha > D$. Moreover, the smaller the initial value of α, the larger the final value: sites grown in more active regions eventually become more strongly screened.

8. Conclusions

Diffusion-limited growth is a beguilingly simple problem: the growth is determined by the geometry. The key to understanding it lies in relating the distribution of growth back to a resulting cluster geometry, thus determining the morphology selected after large times of growth. What we have shown in this paper is that for some purposes it suffices to restrict attention to the foremost tips, whose growth the rest of the cluster must follow. Even our results for screening are consistent with this: the geometrical interpretation of these is that the microscopic tip at which growth occurs evolves by coarsening at successively larger scales, with just a crossover scale between the structure of exposed growth and fully screened interior.

References

Ball, R C 1986 *Physica* **140** A, 62–69
Ball, R. C. 1989 In *Proceedings of Special Seminar on Fractals, Erice 1988* (In the press.)
Ball. R. C. & Blunt, M. J. 1989a *Phys Rev* A. (In the press.)
Ball. R. C. & Blunt. M J 1989b (In preparation.)
Ball, R. C. & Brady, R. M 1984 *J Phys.* A **18**, 809–813.
Ball, R. C., Brady, R. M., Rossi, G & Thompson, B. R. 1985 *Phys. Rev. Lett.* **55**, 1406–1409
Ball, R. C., Nauenberg, M & Witten, T. A 1984 *Phys. Rev.* A **29**, 2017–2020.
Ball, R. C. & Rath Spivack, O 1989 (In preparation.)
Ball, R. C. & Witten, T A 1984 *Phys. Rev.* A **29**, 2966–2967.
Halsey, T. C., Meakin, P. & Procaccia, I. 1986 *Phys Rev. Lett.* **56**, 854–857.
Halsey. T. C. 1987 *Phys Rev Lett* **59**. 2067–2070.
Huang, S.-C. & Glicksman. M. E 1981 *Acta metall.* **29**, 701, 717.
Kertesz, J. & Vicsek, T. 1986 *J. Phys.* A **19**, L257–262.
Langer, J. S. 1980 *Rev mod. Phys.* **52**, 1–28.
Mandelbrot, B. B. 1982 *The fractal geometry of nature.* San Francisco Freeman.
Meakin, P , Majid, I., Havlin, S. & Stanley H. E. 1984 *J Phys* A **17**, L975–L981.
Meakin, P. 1987a *Faraday Discuss. chem. Soc.* (Preprint.)
Meakin, P. 1987b *Phys. Rev.* A **36**, 332–339.
Meakin, P. & Eckmann. J.-P. 1989 (Preprint.)
Meakin, P. 1989 In *Proceedings of Special Seminar on Fractals. Erice 1988.* (In the press.)
Mullins, W. W. & Sekerka, R. F. 1963a *J. appl. Phys.* **34**, 323–329.
Mullins, W. W. & Sekerka, R. F. 1963b *J. appl. Phys.* **35**, 444–451
Muthukumar, M. 1983 *Phys. Rev. Lett* **50**, 839–842
Niemeyer, L., Pietronero, L & Wiesmann, H. J. 1984 *Phys. Rev. Lett.* **52**, 1033–1036.
Nittman, J. & Stanley, H. E. 1986 *Nature, Lond.* **321**, 663–668.
Shraiman, B & Bensimon, D. 1984 *Phys. Rev.* A **30**, 2840–2842
Tang, C. 1985 *Phys. Rev.* A **31**, 1977, 1979.
Turkevich, L. & Scher, H. 1985 *Phys. Rev. Lett.* **55**, 1026–1029
Witten, T. A. & Sander, L M. 1981 *Phys. Rev. Lett.* **47**, 1400–1403.

Discussion

J. S. ROWLINSON (*Physical Chemistry Laboratory, University of Oxford, U.K.*). Dr Ball has shown that clusters grown on two-dimensional lattices under conditions of diffusion-limited aggregation reflect the symmetry of the lattice if this has only a few nearest-neighbour sites. He says that he believes that this lack of circular symmetry will vanish if the number of neighbours is greater than about six. Is there good evidence for such a finite limit, or does the asymmetry simply vanish smoothly as the number increases?

R. C. BALL. Conformal mapping arguments (Ball 1986) suggest that in two dimensions the maximum number n_{max} of stable fingers is given by

$$(\tfrac{1}{2}n_{max} - 1)(D-1) = 1,$$

i.e. $n_{max} = 6$ for the lowest possible value of $D = \tfrac{3}{2}$.

What is less clear is what happens when all the possible finger directions cannot be sustained: is an orderly subset selected spontaneously or do we approach the statistically isotropic limit? This is an open question.

Diffusion-limited aggregation

By P. Meakin and Susan Tolman

Central Research and Development Department,
Experimental Station, E. I. du Pont de Nemours and Company,
Wilmington, Delaware 19898, *U.S.A.*

Improved algorithms have been developed for both off-lattice and hypercubic lattice diffusion-limited aggregation (DLA) in dimensionalities (d) 3–8 and for two-dimensional off-lattice DLA. In two-dimensional off-lattice DLA a fractal dimensionality (D) of about 1.71 was obtained for clusters containing up to 10^6 particles. This is significantly larger than the value of $(d^2+1)/(d+1)$ ($\frac{5}{3}$ for $d=2$) predicted by mean field theories. For $d>2$ the off-lattice simulations give results that are consistent with the mean field theories. For $d=3$ and $d=4$ the effects of lattice anisotropy can easily be seen for clusters containing 3×10^6 and 10^6 sites respectively and the effective fractal dimensionalities are slightly smaller for the lattice model clusters than for the off-lattice clusters. Results are also presented for two-, three- and four-dimensional lattice model clusters with noise reduction.

Introduction

Pattern-formation processes have been of considerable scientific interest and practical importance for many decades. Interest in the growth of complex structures under non-equilibrium conditions has been stimulated by several recent developments. The dissemination of the concepts of fractal geometry (Mandelbrot 1982) and related ideas have provided us with ways of describing a very broad range of irregular structures in quantitative terms. It has been shown that even very simple nonlinear systems and models (see, for example, May 1976; Lorenz 1963; Feigenbaum 1978) can lead to complex, often chaotic, behaviour that can frequently be described in terms of fractal geometry. In addition, it has been shown that simple models for growth and aggregation processes frequently lead to disorderly structures that exhibit a spatially chaotic fractal geometry. In particular, the introduction of the diffusion-limited aggregation (DLA) model (Witten & Sander 1981) led to a renewed interest in models for growth, aggregation and morphogenesis. The increased power, availability and ease of use of digital computers played a crucial role in all of these developments.

In the DLA models, particles are added, one at a time, to a growing cluster or aggregate of particles via random-walk trajectories originating from outside of the region occupied by the cluster. This simple process leads to the formation of a random fractal structure with a fractal dimensionality that is substantially smaller than that of the space or lattice in which the aggregate is embedded. Although the name diffusion-limited aggregation is now well accepted, the DLA model does not describe growth from a finite-density field obeying the diffusion equation. Instead, it describes a random growth process in which the growth probabilities at the surface of the growing cluster are controlled by a scalar field ϕ that obeys the Laplace equation $\nabla^2\phi = 0$. The random walkers are used here to simulate the scalar field ϕ that obeys the Laplace equation with the boundary conditions $\phi = 0$ on the cluster and $\phi = 1$ at infinity. The close relation between the DLA process and growth controlled by a laplacian or harmonic field is made explicit in the dielectric breakdown model (Niemcyer *et al.* 1984).

The DLA model is important because it provides a basis for understanding a wide variety of phenomena such as electrodeposition, random dendritic growth, fluid–fluid displacement in Hele–Shaw cells and porous media, dissolution of porous media, dielectric breakdown and possibly biological processes such as the growth of nerve cells and blood vessels (for reviews see Ball 1986a; Meakin 1988; Matsushita 1989). Despite the apparent simplicity of the DLA model, it provides an important theoretical challenge that has not yet been fully met. Despite the introduction of several promising new approaches (Turkevich & Scher 1985; Ball *et al.* 1985; Halsey *et al.* 1986; Ball 1986b; Procaccia & Zeitak 1988; Bohr *et al.* 1989) we do not yet have a completely satisfactory theory for DLA. In fact, at the present time, there is still substantial uncertainty concerning even the correct qualitative description of DLA clusters generated by computer simulations and these uncertainties are not likely to be reduced in the absence of a better theoretical understanding.

Despite the practical difficulties of determining the asymptotic scaling behaviour of DLA clusters from finite (often quite small-scale) simulations, much of our knowledge and understanding of DLA still comes directly from computer simulations. The main purpose of this paper is to describe some recent advances in this area with an emphasis on simulations using spaces and lattices with euclidean dimensionalities $(d) \geq 3$, a subject that has been neglected in the past.

The DLA model

A simple two-dimensional square lattice diffusion-limited aggregation model is illustrated in figure 1. This figure shows a small cluster with a maximum radius of r_{max} measured from the original seed or growth site. Particles are launched one at a time from a randomly selected position on a 'launching' circle of radius $r_{max} + 5$ lattice units and follow random walk trajectories on the lattice. The particle may follow a trajectory (like t_1 in the figure) that eventually brings it to an unoccupied perimeter site. In this event the perimeter site is filled and the cluster grows. Alternatively, the particle may follow a trajectory (like t_2) that eventually moves it a long distance from the cluster (a distance of kr_{max}, where k is 3 in figure 1). If this happens, the trajectory is terminated and a new random-walk trajectory is started from a randomly selected position on the launching circle. This procedure is repeated many times until a large cluster has been generated. Algorithms similar to this can be used to generate clusters containing a few thousand sites or particles (Witten & Sander 1981).

The efficiency of the DLA algorithm can be dramatically improved by allowing the random walker to take long steps when it is far from any of the sites occupied by the cluster (Ball & Brady 1985; Meakin 1985). If the random walker is enclosed

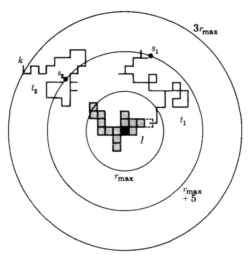

F<small>IGURE</small> 1. A simple model for two-dimensional square-lattice D<small>LA</small>. The filled square represents the original seed or growth site, the shaded squares represent sites that have already been filled by the growth process and the open site with broken border is an unoccupied perimeter site that is entered by the random walker following trajectory t_1.

in a region with no occupied (absorbing) sites or particles and with a simple shape (at the centre of a circle or square for example), we can replace the random walk within that region by a single step to an appropriately selected position at its edge. For example, if the random walker is the centre of an empty circle, we can immediately transfer the particle to a randomly selected position on the circle. To take advantage of the ensuing reduction in the number of steps in the random walk, we must have an efficient way of determining the distance from the random walker to nearest occupied site(s) on the cluster (Ball & Brady 1985; Meakin 1985). For two-dimensional clusters containing about 10^5 particles or sites this procedure reduced the computer time needed to grow a cluster by a factor of about 1000. In addition, the distance kr_{max} at which trajectories are terminated can be made very large (100 r_{max} is typical of recent algorithms) thus eliminating any bias that may arise by terminating the trajectories too close to the cluster. Clusters containing more than about 10^5 particles or sites cannot be represented by filled and empty sites on a square lattice. Instead, the cluster is represented by a hierarchy of 'maps' on different length scales (L_n) that are updated as needed as the cluster grows. The nth level map consists of a lattice whose elements represent regions of size $L_n \times L_n$. If the lattice element is 'empty' or 'off', then a random walker in this region can take a step of length L_n. If the lattice element is 'full' or 'on', then the more detailed map at level $n-1$ is examined to determine if the walker may take a step of length $L_{n-1} = \frac{1}{2}L_n$. The maps on shorter and shorter length scales are examined until the random walker is allowed to make a move or until the lowest level is reached. At the lowest (most detailed) level each element of the map contains information about the exact location of sites or particles that can be used to determine if the walker has contacted the cluster or if a step of the minimum length (one lattice unit for lattice models or a distance on the order of the particle radius for off-lattice models) should be made. To reduce information storage requirements only those portions of the lower-level, more detailed maps that are needed (regions close to the cluster) are constructed and updated. Figures 2 and 3 show two-dimensional square lattice and off-lattice clusters generated by

30 000 lattice units

FIGURE 2. A 1.27×10^7 site square DLA cluster grown by using the algorithm of Ball & Brady (1985).

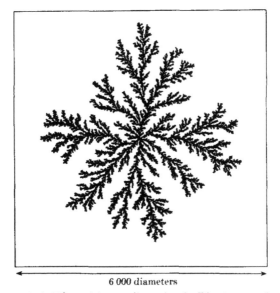

6 000 diameters

FIGURE 3. A 10^6 particle two-dimensional off-lattice DLA cluster.

using algorithms of this type. Figure 2 shows a 1.27×10^7 site square-lattice DLA cluster grown by using the algorithm of Ball & Brady (1985). In this case the effects of the weak lattice anisotropy on the overall shape of the cluster can easily be seen. Figure 3 shows a 10^6 particle off-lattice DLA cluster (Tolman & Meakin 1989). Similar algorithms have been developed for DLA in hypercubic lattices and spaces with dimensionalities of 3–8 (Tolman & Meakin 1989). The details concerning the structure of the hierarchy of maps and the way information is stored at the lowest level vary with d and the size of the clusters that are required. However, the general structure of the algorithms is very similar to that described above.

RESULTS

Two-dimensional DLA

Computer simulations have been used quite extensively to investigate the structure of both lattice and off-lattice models for DLA. Much of this work has been reviewed quite recently (Meakin 1988) and the structure of square-lattice DLA clusters containing up to 4×10^6 sites has been investigated (Meakin *et al.* 1987). Here some more recent results are presented.

A quite large number (221) of 10^6 particle off-lattice DLA clusters were grown and the radius of gyration (R_g) of each of the clusters was determined for each 5% increment in the cluster size (s). The dependence of $\ln(R_g)$ on $\ln(s)$ is quite linear for clusters containing more than a few hundred particles and it is apparent that the dependence of R_g on s can be described very well by the power law

$$R_g \sim s^\beta. \tag{1}$$

The corresponding effective fractal dimensionality $D_\beta (D_\beta = 1/\beta)$ was obtained by least-squares fitting the coordinates $(\ln(R_g), \ln(s))$ by a straight line for clusters in the size range $s_1 \leqslant s \leqslant s_2 (s_2 = 1.05^9 s_1)$. Figure 4 shows the dependence of the exponent β on $s ((s_1 s_2)^{\frac{1}{2}})$ obtained in this way.

It is apparent that β is essentially independent of s. By least-squares fitting the dependence of R_g on s for clusters in the size range $10^5 \leqslant s \leqslant 10^6$ a value of 0.5837 ± 0.0014 was obtained for β corresponding to a fractal dimensionality of

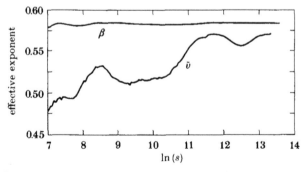

FIGURE 4. Dependence of the exponents β and $\bar{v}$, which describe the growth of the radius of gyration and width of the active zone on the cluster size s. The results for β and $\bar{v}$ were obtained from 221 and 93 10^6 particle off-lattice two-dimensional DLA clusters respectively.

1.713 ± 0.004. This result is in excellent agreement with that obtained earlier ($\beta \simeq 0.584$; Meakin & Sander 1985, and unpublished results). It appears that this is by far the most accurately known exponent associated with any DLA model and there is no evidence that the asymptotic value for the fractal dimensionality D_β is different from that obtained from quite small clusters for two-dimensional off-lattice DLA.

For some of the 10^6 particle clusters we also measured the width of the active zone, ξ (variance in the deposition radius; Plischke & Racz 1984) as a function of s. The dependence of ξ on s can be represented by the power law

$$\xi \sim s^{\bar{v}}. \tag{2}$$

Using quite small square-lattice DLA clusters, Plischke & Racz (1984) found an effective value of about 0.484 for $\bar{v}$. However, the effective value for $\bar{v}$ increases with increasing cluster size and reaches a value of about 0.54 for off-lattice clusters in the size range 25000–50000 particles. This suggests that in the asymptotic ($s \to \infty$) limit $\bar{v}$ might be equal to β (Meakin & Sander 1985). Figure 4 also shows the results obtained for $\bar{v}$ from 93 10^6 particle off-lattice clusters. It is apparent that $\bar{v}$ continues to increase with increasing cluster size and reaches a value quite close to that found for β for clusters containing 10^6 particles.

We have also investigated the cluster-size distribution (N_s) for the incremental growth in DLA clusters. The clusters are grown to a size of s_1 sites or particles and then an additional s_2 particle is added. The quantity N_s is then the number of clusters of size s in the incremental growth consisting of the last s_2 particles added. We might expect (Racz & Vicsek 1983; Matsushita & Meakin 1988) that the dependence of N_s on s should be a power law

$$N_s \sim s^{-\tau} \tag{3}$$

with τ given by
$$\tau = 1 + D_1/D, \tag{4}$$

where D_1 is the dimensionality of the old growth/new growth interface (Meakin & Witten 1983). Figure 5 shows the results obtained from 104 10^5 particle off-lattice DLA clusters for $s_1 + s_2 = 10^5$ and three different values for s_1 (25000, 50000 and 75000). It is apparent from figure 5 that the dependence of N_s on s is not really algebraic. This may be a consequence of the fact that the old growth/new growth interface is not fully saturated and there are a lot of small clusters that will grow as s_2 is further increased. However, the results shown in figure 5 do suggest that τ might have a value of about 1.68. It has been argued that the fractal dimensionality of the old growth/new growth interface should be 1.0 and in this event we would expect to obtain a value of $1 + \beta$ or about 1.58 for τ. A value of 1.68 indicates that D_1 should have a value of about 1.16, which is in good agreement with simulation results (a value of 1.13 was found for D_1 by Meakin & Witten (1983) using square-lattice DLA clusters). Results very similar to those shown in figure 5 were obtained from 435 10^5 site square-lattice DLA clusters.

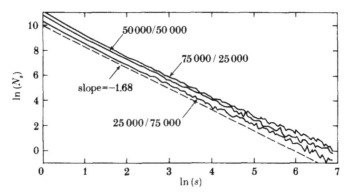

FIGURE 5. The cluster-size distributions N_s for the incremental growth obtained from 104 10^5 particle off-lattice DLA clusters. The three curves show the results obtained for $s_1 = 25000$, 50000 and 75000 and $s_2 = 75000$, 50000 and 25000 respectively, where s_1 is the number of particles in the old growth and s_2 is the number of particles in the new or incremental growth.

Three-dimensional DLA

Relatively few studies of the structure of three-dimensional DLA clusters have been carried out. An effective fractal dimensionality (D_β) of about 2.49 has been obtained from approximately 100 50000 site cubic lattice DLA clusters (P. Meakin & L. M. Sander, unpublished work). Figure 6 shows a projection of a 3×10^6 site cubic-lattice DLA cluster work onto a plane and a cross section through the origin of the cluster along a parallel plane. The effects of the weak cubic-lattice anisotropy (that are not readily apparent in clusters containing 10^4 sites) are quite evident in

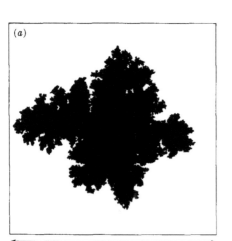

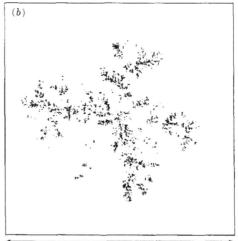

1 100 lattice units   1 100 lattice units

FIGURE 6. A projection (a) and a cross section through the origin (b) for a 3×10^6 site cubic-lattice DLA cluster.

this figure. The overall dimension (linear size) of this cluster is the same as that for square-lattice DLA clusters containing about 5×10^4 sites. It is at about this size that square-lattice DLA clusters have a diamond-like shape which eventually evolves into a cross-like shape (figure 1). Because a quite large amount of computer time (about 10 h of CPU time on an IBM 3090 computer) is required to grow 3×10^6 site clusters, more quantitative results were obtained from 98 1 250 000 site cubic-lattice DLA clusters, 138 300 000 site clusters, 482 100 000 site clusters and 169 100 000 particle three-dimensional off-lattice DLA clusters. Figure 7 shows the cluster-size dependence of the effective fractal dimensionality D_β obtained from the 100 000 particle off-lattice clusters and the 1 250 000 site cubic-lattice clusters. For the off-lattice clusters the fractal dimensionality is very close to the value of 2.50 $((d^2+1)/(d+1))$ obtained from mean field theories (Muthukumar 1983; Tokuyama & Kawasaki 1984; Matsushita *et al.* 1986). For the cubic-lattice model D_β has an effective value slightly smaller than 2.50 (about 2.48 for 10^6 site clusters). This is presumably a consequence of lattice anisotropy and similar effects have been seen in square-lattice DLA (Meakin 1986; Meakin *et al.* 1987) but it appears from figure 7 that enormous clusters would be required to see an effective fractal dimensionality significantly smaller than 2.48.

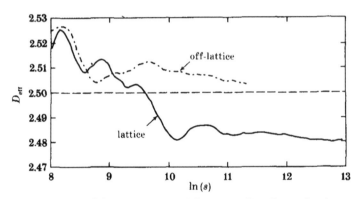

FIGURE 7. Dependence of the effective fractal dimensionality (D_β) on the cluster size s obtained from off-lattice and cubic-lattice DLA clusters. The broken line indicates the mean field theory value of 2.50.

We have also measured the width of the active zone (ξ) for the three-dimensional DLA clusters. For the off-lattice model $\bar{v}$ increases from a value of about 0.31 for clusters containing a few thousand particles (a result in good agreement with that obtained by Racz & Plischke (1985) from cubic-lattice DLA clusters of the same size) to a value of about 0.34 for clusters containing about 10^5 particles. For cubic-lattice DLA the exponent $\bar{v}$ continues to increase with increasing s above 10^5 sites, but in this case the contribution of the growing cluster anisotropy apparent in figure 6 may be important.

Figure 8 shows the cluster size distribution in the incremental growth obtained from 138 300 000 site cubic-lattice clusters for old growth/new growth sizes of 75 000/225 000, 150 000/150 000 and 225 000/75 000 sites. Similar results were

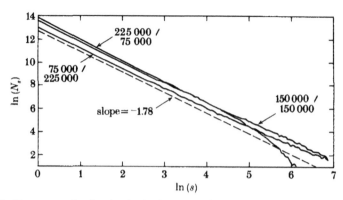

FIGURE 8 Cluster-size distribution in the incremental growth obtained from 300000 site cubic-lattice DLA clusters. The sizes of the old growth and the new growth or incremental growth (s_1/s_2) are shown by each of the curves.

obtained from larger (6×10^5 sites) and smaller (10^5 sites) clusters and from off-lattice simulations. The cluster-size distributions in the incremental growth cannot be described accurately by equation (3) (with a cut-off at large sizes). However, it is apparent from figure 8 that the size distribution can be described approximately by equation (3) with an exponent (τ) of about 1.8. This would correspond to an interface dimensionality of about 2.0. This result is in accord with a direct measurement of D_i (Meakin & Witten 1983) with relatively small cubic-lattice clusters.

DLA in dimensionalities greater than three

Clusters containing about 10^5 particles or sites have been grown for $d = 4$–6 by using both off-lattice and hypercubic-lattice models. Figure 9 shows a projection and a cross section for a 10^6 site four-dimensional hypercubic-lattice model cluster. The effects of the lattice anisotropy can be clearly seen in figure 9a. b. For all of these models the dependence of the cluster radii of gyration on cluster size has been measured to obtain the effective fractal dimensionality D_β. For $d = 2$–5 the effective value for D_β is smaller than the 'mean field' value of $(d^2 + 1)/(d + 1)$ for small clusters. As the cluster size increases, D_β increases and exceeds the mean field value. As the cluster continues to grow, D_β decreases and eventually reaches a value very close to the mean field value. For lattice models a substantially smaller effective fractal dimensionality can be seen for large two-dimensional clusters and a slightly smaller value for $d = 3$ and 4 is reached at the largest attainable cluster sizes. Figure 10 shows the dependence of D_β on s obtained from 82 five-dimensional off-lattice clusters and 241 five-dimensional hypercubic-lattice model clusters each containing 10^5 particles or sites. In this case the effective fractal dimensionality for the largest cluster sizes is quite close to the mean field value of 4.333 for both the lattice and off-lattice models. It appears that for $d \geqslant 5$ clusters containing 10^5 or fewer sites are in the fluctuation-dominated régime and that the hypercubic-lattice anisotropy has little effect on the structure of clusters of this size.

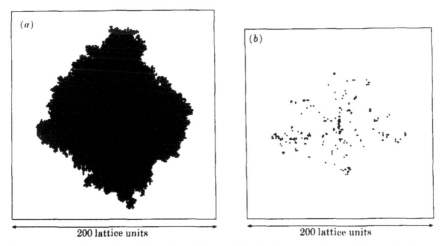

200 lattice units   200 lattice units

FIGURE 9. A 10^6 site four-dimensional hypercubic-lattice DLA cluster. Figure 9a shows a projection of the cluster onto a plane and (b) shows a cross section through the cluster that intersects the original seed or growth site.

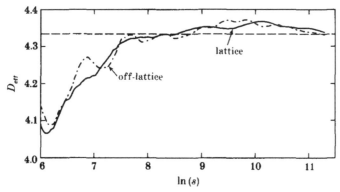

FIGURE 10. The cluster-size dependence of the effective fractal dimensionality D_β obtained from 100000 sites or particles five-dimensional hypercubic-lattice and off-lattice DLA clusters. The mean field theory value of 4.333 is indicated by a broken horizontal line.

For $d \geqslant 6$ we appear to see only the first part of the dependence of D_β on s. D_β increases to a value that exceeds the mean field value of $(d^2+1)/(d+1)$ but does not decrease. Figure 11 shows an example of this behaviour obtained from 28 70000 site seven-dimensional hypercubic-lattice DLA clusters. It is not surprising that the asymptotic behaviour is not seen in this case because a cluster of 70000 sites has a radius of gyration of only about 6.5 lattice units and a maximum radius of about twice that. For $d \geqslant 6$ the dependence of D_β on s is the same (within the statistical uncertainties of our simulations) for both lattice and off-lattice models.

The cluster-size distribution has been measured for the incremental growth in both the four- and five-dimensional models. Values of about 1.70 and 1.65

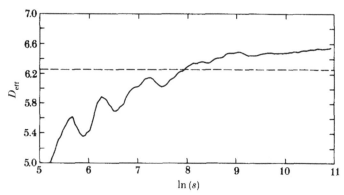

FIGURE 11. Cluster-size dependence of the effective fractal dimensionality obtained from seven-dimensional hypercubic-lattice DLA clusters. The horizontal line indicates the fractal dimensionality of 6.25 predicted by the mean field theory.

respectively were obtained for the exponent τ, which describes the cluster-size distribution for the incremental growth. With equation (4) these results indicate an interface dimensionality (D_i) of about 2.4 for $d = 4$ and about 2.8 for $d = 5$. Because of the small spatial extent (overall size) of these clusters, the uncertainties associated with these values are quite large.

Noise-reduced DLA

In the noise-reduced DLA model (Tang 1985; Kertesz & Vicsek 1985) random walkers are used to contact the surface of a growing cluster by using the DLA

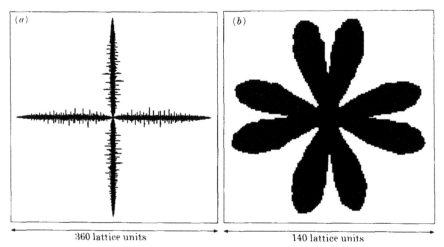

FIGURE 12. Clusters grown with random walkers and a noise-reduction parameter (m) of 10 000 with DLA boundary conditions (a) and dielectric breakdown model boundary conditions (b). Both clusters eventually evolve into complex patterns, but for (a) this occurs via side branching and for (b) via tip splitting.

algorithm. After a random walker has contacted an unoccupied perimeter site, a record of the contact is kept (the contact score associated with that site is incremented by one) and a new random walker is started from a randomly selected position on the launching circle. Only after a perimeter site has been contacted m times in this fashion does it become occupied. After a growth event all of the new perimeter sites are given a score of zero, but the old perimeter sites retain their contact scores, which continue to accumulate. In this model $m = 1$ corresponds to ordinary DLA. Clusters grown in this fashion with small values of m strongly resemble much larger clusters grown with $m = 1$ (figure 1) (Kertesz et al. 1986). It is widely believed, but has never been rigorously shown, that the structure of small clusters grown with noise reduction faithfully represents the structure of much

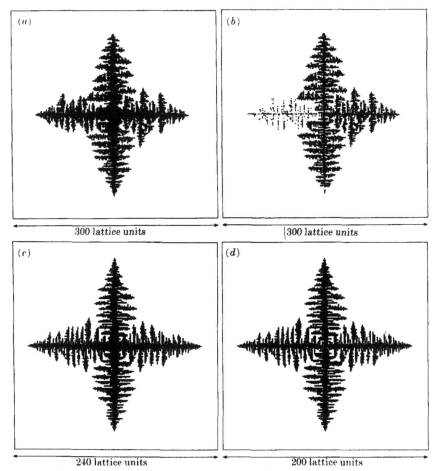

300 lattice units   300 lattice units

240 lattice units   200 lattice units

FIGURE 13. Noise-reduced cubic-lattice DLA clusters. (a, b) Projection and a cross section for a 43 253 site cluster grown with a noise reduction parameter of 30. In (b) the cross section is through the cluster origin and one of the cluster arms (on the left-hand side) deviates from this plane. (c, d) A projection and cross section for a 28 350 site cluster grown with a noise-reduction parameter (m) of 100.

larger clusters grown without noise reduction and that the asymptotic behaviour of lattice models for DLA can be investigated by using noise reduction.

In the dielectric breakdown model the growth probability P_i associated with the ith unoccupied perimeter site is given by

$$P_i \sim n\phi_i, \tag{5}$$

where ϕ_i is the potential (obtained by solving the discretized Laplace equation) at the ith perimeter site and n is the number of occupied nearest neighbours. This model can be implemented with random walkers by allowing the walker to step onto the cluster and occupy the previously visited unoccupied perimeter site (L. Pietronero, personal communication 1985). Ordinarily the differences between the DLA and dielectric breakdown model clusters that result from the different local boundary conditions are quite small. However, for noise-reduced DLA and related models (Nittmann & Stanley 1985) these local boundary conditions become more important (R. C. Ball, personal communication 1987; L. Pietronero, personal communication 1987). This is illustrated in figure 12, which shows clusters grown by using both models (with random walkers) with a noise reduction parameter (m) of 10000. Noise-reduced DLA clusters first grow as compact crosses that begin to side branch when the cluster size s reaches a value proportional to $(\log(m))^3$ (Eckmann *et al.* 1989). The noise-reduced DLA cluster shown in figure 12a has just passed this stage. The dielectric breakdown model cluster on the other hand (figure 12b) grows four arms that undergo tip splitting rather than side branching.

We have also done noise-reduced DLA simulations with three-dimensional cubic and four-dimensional hypercubic lattices. Figure 13 shows a projection and cross section for a 43253 site cubic-lattice cluster grown with a noise-reduction parameter (m) of 30 and for a 28350 site cluster grown with a noise-reduction parameter of 100. The structures generated by these models resemble quite closely those generated by the disorderly growth of cubic crystals under non-equilibrium conditions (Garcia-Ruiz 1986).

Figure 14 shows the dependence of the cluster radius of gyration on the cluster size plotted as $\ln(R_g/s^{\beta^*})$ against $\ln(s)$ where β^* is the mean field radius of gyration

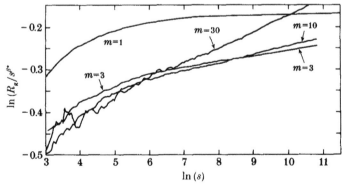

FIGURE 14 Dependence of $\ln(R_g/s^{\beta^*})$ on $\ln(s)$ obtained from four-dimensional DLA clusters with noise-reduction parameters of $m = 1$ (ordinary DLA), 3, 10 and 30. Here β^* is the mean field radius of gyration exponent $(\beta^* = (d+1)/(d^2+1) = \frac{5}{17})$.

exponent $(\frac{5}{17})$. The dependence of $\ln(R_g/\beta^*)$ on $\ln(s)$ seems quite linear (in contrast to the behaviour found for $d = 2$ where strong oscillations are seen). For the largest value of m ($m = 30$) the effective fractal dimensionality obtained from the dependence of R_g on s is about 3.01.

CONCLUSIONS

Since the DLA model was introduced about seven years ago by Witten & Sander (1981), the size of clusters that we are able to grow and the speed with which they can grow has increased substantially. This has allowed us to reduce uncertainties caused by both statistical and finite size effects. Part of this improvement has come about as a result of much larger amounts of computer time on much faster computers. However, the major contribution has come from improved algorithms. It seems unlikely that the speed, information storage capability and availability of computers will increase sufficiently during the next 5–10 years to allow us to approach significantly closer to the asymptotic limit and thus resolve many of the outstanding questions concerning both off-lattice and lattice models for DLA. Based on the results reported here, it seems that truly enormous clusters will be required to see the asymptotic effects of lattice anisotropy. It is also probable that the algorithms used to grow DLA clusters will improve at a slower rate during the next five years than they did during the past five years. Clearly a less 'brute force' approach will be needed if further progress is to be made. The use of strategies like noise reduction to see the asymptotic behaviour of DLA could be extremely valuable. However, a better theoretical understanding is needed to use this approach with confidence.

The use of models closely related to DLA to explore non-equilibrium growth processes is still a rapidly growing area. Models of this type can be of considerable value in developing a better understanding of specific physical or chemical processes. Unfortunately, some of the work in this area seems to be poorly motivated and to have little to do with physical reality.

A complete theoretical understanding of DLA still eludes us. A satisfactory theory for a two-dimensional noise-reduced DLA (Eckmann et al. 1989) has been developed that leads to quite detailed predictions concerning the cluster structures that have been verified by using computer simulation. Although this is an encouraging development, it does not provide us with an understanding of DLA in the fluctuation dominated or off-lattice régime.

Because of the importance of DLA as a paradigm for non-equilibrium growth and the theoretical challenge posed by this simple-to-define model, it is likely to be the subject of considerable theoretical attention during the next few years. The simulation results presented above for the fractal dimensionality D_β are consistent with the predictions of several mean theories $(D = (d^2 + 1)/(d + 1))$ in the off-lattice or fluctuation dominated régime for $d > 2$. However, for $d = 2$ there is a clear discrepancy between the mean field predictions and the simulation results. If the mean theoretical result is correct for $D \geqslant 3$, a more rigorous justification is needed.

The work described in this paper was stimulated by discussions with many colleagues and collaborators. We particularly thank R. C. Ball for many valuable ideas and for making the Ball–Brady (1985) algorithm for square-lattice DLA available to us. The work on incremental-growth cluster-size distributions was carried out in collaboration with M. Matsushita.

REFERENCES

Ball, R. C 1986a In *On growth and form · fractal and non-fractal patterns in physics* (ed. H. E Stanley & N. Ostrowsky), NATO ASI Series E100. p. 69. Dordrecht: Martinus Nijhoff.

Ball, R. C. 1986b *Physica* **140** A, 62.

Ball, R C. & Brady, R. M 1985 *J. Phys.* A **18**, L809.

Ball, R. C., Brady, R M , Rossi, G. & Thompson, B. R. 1985 *Phys. Rev. Lett.* **55**, 1406.

Bohr, T , Cvitanovic, P & Jensen, M. H 1989 (Preprint)

Eckmann, J. P , Meakin, P , Procaccia, I. & Zeitak, R. 1989 (Preprint.)

Feigenbaum, M J 1978 *J statist. Phys* **19**, 25.

Garcia-Ruiz, J. M. 1986 *J. Cryst. Growth* **75**, 441.

Halsey, T. C.. Meakin, P & Procaccia, I. 1986 *Phys. Rev. Lett.* **56**, 854.

Kertesz, J & Vicsek, T. 1985 *J Phys.* A **19**, L257

Kertesz, J., Vicsek, T. & Meakin, P. 1986 *Phys. Rev Lett.* **57**, 3303.

Lorenz, E. N 1963 *J atmos Sci.* **20**, 130

Mandelbrot. B. B. 1982 *The fractal geometry of nature.* New York: W. H. Freeman and Company.

Matsushita, M. 1989 In *The fractal approach to heterogeneous chemistry. surfaces, colloids polymers* (ed. D Avnir). New York John Wiley and Sons

Matsushita, M.. Honda, K., Toyoki, H., Hayakawa, Y & Kondo, H. 1986 *J. phys. Soc. Japan* **55**, 2618.

Matsushita, M. & Meakin, P. 1988 *Phys Rev* A **37**, 3645

May, R M. 1976 *Nature, Lond.* **261**, 459

Meakin, P. 1985 *J. Phys.* A **18**. L661

Meakin, P. 1986 *Phys. Rev.* A **33**, 3371.

Meakin, P. 1988 In *Phase transitions and critical phenomena* (ed. C. Domb & J. L. Lebowitz), vol. 12, p. 335. New York: Academic Press.

Meakin, P., Ball, R. C., Ramanlal, P. & Sander, L M. 1987 *Phys. Rev.* A **35**, 5233.

Meakin, P & Sander, L M. 1985 *Phys. Rev. Lett.* **54**, 2053.

Meakin, P. & Witten, T. A. 1983 *Phys. Rev* A **28**, 2985.

Muthukumar, M 1983 *Phys. Rev. Lett* **50**, 839.

Niemeyer, L., Pietronero, L. & Wiesmann, H. J. 1984 *Phys. Rev. Lett.* **52**, 1033.

Nittmann, J. & Stanley, H. E. 1985 *Nature, Lond* **321**, 663.

Plischke, M. & Racz, Z 1984 *Phys Rev. Lett.* **53**, 415

Procaccia, I & Zeitak. R. 1988 *Phys Rev Lett.* **60**, 2511.

Racz. Z. & Plischke, M 1985 *Phys. Rev.* A **31**. 985.

Tang. C. 1985 *Phys. Rev.* A **31**, 1977

Tokuyama, M & Kawasaki, K. 1984 *Phys Lett.* **100** A, 337

Tolman. S. & Meakin, P. 1989 (Preprint.)

Turkevich, L. A & Scher, H 1985 *Phys. Rev. Lett.* **55**, 1026

Witten, T. A & Sander, L M. 1981 *Phys. Rev. Lett.* **47**, 1400.

Discussion

A. BLUMEN (*University of Bayreuth, F.R.G.*). Is there a transparent argument to understand that a weak anisotropy on a microscopic scale, such as an underlying lattice, leads to such prominent effects (diamond-shaped DLA) on the macroscopic scale?

P. MEAKIN. Early, small-scale simulations of DLA on a square lattice provided no obvious indications of anisotropy. When larger-scale simulations (Ball & Brady 1985) indicated that lattice anisotropy had an important effect on the overall cluster geometry, this came as a surprise. A recent theoretical analysis of noise-reduced DLA with anisotropy (Eckmann *et al.* 1989) provides an understanding of the shape of DLA clusters with large amounts of noise reduction. If it is accepted that noise reduced DLA reflects the asymptotic (large-size) structure of ordinary square lattice DLA, then the cross-like structure of large DLA clusters (figure 2) seems reasonable. In the noise-reduced DLA model, each site in the cluster can be considered to represent a region of a DLA cluster containing many sites (i.e. the growth process has been 'coarse-grained'). As the noise-reduction parameter (m) is increased, each lattice in the noise-reduced cluster represents a larger and larger region in the ordinary DLA cluster. However, there is no firm theoretical foundation for the correspondence between noise-reduced DLA and square lattice DLA and we do not have a quantitative relation between the noise-reduction parameter (m) and the scale of coarse graining. At present, there is no theoretical understanding (or quantitative simulation results) of how the effect of the weak lattice anisotropy grows as the cluster size increases. However, simulations addressing this question are in progress.

The growth of the anisotropy in lattice models for DLA can probably be understood in terms of the Mullins–Sekerka instability (Mullins & Sekerka 1963). Ball (1986a) has discussed how the Mullins–Sekerka instability is modified for the growth of symmetrically arranged arms. This analysis indicates that four arms should be stable and is consistent with the asymptotic shape of square lattice DLA.

Reference

Mullins, W. W. & Sekerka, R. F. 1963 *J. appl. Phys.* **34**, 323.

Electrodeposition in support: concentration gradients, an ohmic model and the genesis of branching fractals

By D. B. Hibbert[1] and J. R. Melrose[2]

[1] Department of Analytical Chemistry, University of New South Wales, P.O. Box 1, Kensington, Sydney 2033, Australia
[2] The Blackett Laboratory, Imperial College, Prince Consort Road, London SW7 2BZ, U.K.

[Plates 1 and 2]

The technique of paper-supported copper electrodeposition provides examples of well-presented fractal and dense radial structures. The growths may be developed to reveal concentration gradients around the growths at low cell overpotential. Measurements for current and length scale against time, within a mid-range of cell overpotentials, fit an ohmic model of the growth conditions. To examine the relation of growth morphology to the micrometre-scale structure, we grew first at one overpotential and then continued at a lower overpotential. Electron microscope observations of this growth reveal a distinct change in microstructure from irregular to dentritic microcrystalline from the high to low potential respectively. The interface between the growths is a distinctive compact granular deposit. The granular deposit is unstable to branching and dendrite growth.

Introduction

Electrodeposition has been of much current interest as an experimentally controllable example of growth governed by diffusion. In the ideal experiment the field controlling transport of the depositing species obeys, in the quasi-stationary approximation, Laplace's equation (Ball 1986). Experimental realizations with this field being the concentration (Brady & Ball 1984; Kaufman et al. 1987) and also an electric field (Matsushita et al. 1984, 1985; Grier et al. 1986; Swada et al. 1986) have been reported. The fractals of the diffusion-limited extreme (Witten & Sander 1983) have been grown (Brady & Ball 1984; Kaufman et al. 1987; Kahanda & Tomkiewicz 1988). This extreme has been extensively studied in computer simulations (Meakin 1987, and references therein).

Transport of the depositing species may involve both diffusion and migration, convection and stirring. The growths show a compact structure at the micrometre scale. This compact structure adds the boundary conditions of activation, lattice anisotropy, surface migration and surface tension to the growth laws. The morphologies of the micrometre scale and the centimetre scale are interrelated via these mechanisms and boundary conditions. Elucidation of these relations is a primary aim of the present research. Such understanding may help deducing mechanism from observed structure and conversely, the control of structure in

[149]

materials technology. The morphology of electrodeposits has been of interest in the context of battery design, electroplating technologies and the production of metal powders (Bockris & Reddy 1973, and references therein).

Fractal geometry (Mandelbrot 1982) has made a significant contribution to these problems. It has introduced powerful and quantifiable concepts to morphological description, identified different growth régimes and given strong justification to the use of the computer simulation of growth employing simple rules. For example, the influence of microscopic anisotropy via the scale invariance exhibited by fractals on macroscopic morphology has been clearly demonstrated in simulation (Meakin 1987; Nittman & Stanley 1986). Other simulations have investigated the influence of activation (Vicsek 1984; Voss & Tomkiewicz 1985).

In the deposition of zinc, the variation of millimetre–centimetre-scale growth morphology with concentration and applied potential has been investigated (Grier et al. 1986; Swada et al. 1986). A fractal régime was found at low concentrations and low applied potentials. A variety of other growth morphologies were found: dense radial, dendritic and needle or stringy growth at extreme applied potentials. (We reserve the word dendritic for regularly branched structures.) Different phase diagrams in the space of concentration and applied potential were reported. However, reference electrodes were not used in these experiments and the anode-to-cathode potential is not a meaningful parameter when concentrations are varied (Kaufman et al. 1987). Grier et al. (1986) reported X-ray and transmission electron microscopy (TEM) analysis of the compact structure. They associated diffusion-limited aggregation (DLA) with the absence of long-range order. The micrometre-scale structure of zinc, in particular the genesis of dendritic growths, has been of some past interest (Diggle et al. 1969; Mansfield & Gilman 1970; Bockris et al. 1973). Faust (1968) examined the dependence of growth morphology with deposition current.

The previous experiments have grown the deposits in solution. However, we are interested in electron microscope examination and subsequent experimentation on the growths (e.g. using the growths as electrodes, probing the active surface area of the growths and AC response). For these reasons we have recently (Hibbert & Melrose 1988) developed techniques for growing the deposits supported in filter paper. The initial experiments have been conducted on copper deposition. The paper prevents convective motion of the solvent and provides good sample preservation. On decreasing the cell overpotential, a change in centimetre-scale morphology from dense radial to open branching growth was observed. Image analysis gave fractal dimensions of the centimetre-scale growths in the range 2.0–1.76 (figure 3 below shows a mixed growth with a dense radial centre and more open exterior). However, inhomogeneous and anisotropic growths were often found.

The paper support affords a unique opportunity to observe concentration gradients around the growths. Some early results of this development technique are given here.

We also report here measurements of growth diameter and current against time for three growths within the mid-range to high range of the applied cell overpotentials. These results fit an ohmic model of the current time characteristic.

Previously it was reported (Hibbert & Melrose 1988) that electron microscope analysis revealed a wide variation of compact micrometre-scale structure that fell roughly into three categories; irregular microcrystalline, dendritic or regular microcrystalline, and granular or mossy. In an effort to probe the dependence of morphology on conditions we report on a growth under mixed conditions: the initial stages grown at 3 V and the later stages grown at 1 V.

Experimental details

A copper ring anode of internal diameter 8 cm, external diameter 9.5 cm and thickness 0.5 mm was clamped on top of a sheet of Whatmann 541 filter paper. The tip of a copper wire was placed on the paper central to the anode to act as the cathode and centre of growth. A reference electrode, a copper wire in the experimental solution, was placed in capillary contact close to the anode (because of the growth the reference cannot be placed close to the growth interface, in deference to the usual arrangement). Growth was made from a 0.75 mol dm^{-3} $CuSO_4/1.0$ mol dm^{-3} H_2SO_4 solution; the paper was in contact with a reservoir of the solution. The position of the reference electrode introduces some problems of control of the overpotential at the interface; the cell overpotentials quoted below are specific to the experimental set up and contain significant contributions from the growth and paper resistance. The acid solutions were found necessary for growth in cellulose paper. These and other details were reported earlier (Hibbert & Melrose 1988).

General scheme

A general scheme for the overpotential dependence of growth morphology in these electrochemical systems is given below. The present experiment is included in square brackets. In the scheme the term 'support' refers to a supporting electrolyte as opposed to the paper mechanical support.

The crossover from activation to diffusion control is governed by the Butler–Volmer equation (Vetter 1967). Under activation control the total current will be limited by the available active surface area.

In the diffusion extreme the current will be determined (if we make a circular model of the growth) by the growth radius. If we assume the presence of a diffusion layer of thickness δ. One finds for an absorbing disc of radius R in the plane that the diffusion current is

$$I/nF = 2\pi D(C_s - C_b)/Ln((R+\delta)/R), \qquad (1)$$

where D is the diffusion constant of the depositing species, C_s denotes the surface concentration, and C_b the bulk concentration at δ. The diffusion limit is $C_s \rightarrow 0$.

The diffusion limit is most easily achieved in the presence of a large excess of indifferent supporting electrolyte (one that does not undergo electrochemical reaction at the applied overpotentials). The experiments of Brady & Ball (1984) were done under these conditions. In the absence of an indifferent supporting electrolyte, or, as is the case in the present experiment, a supporting electrolyte

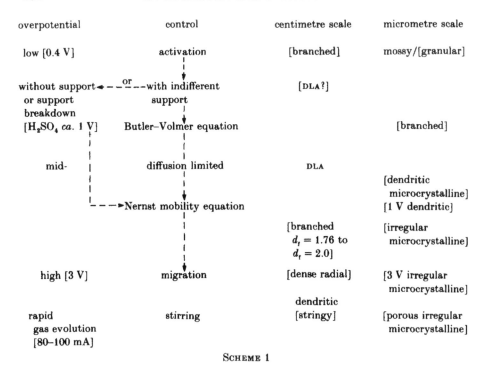

overpotential	control	centimetre scale	micrometre scale
low [0.4 V]	activation	[branched]	mossy/[granular]
without support or support breakdown [H₂SO₄ *ca.* 1 V]	—or— with indifferent support Butler–Volmer equation	[DLA?]	[branched]
mid-	diffusion limited	DLA	[dendritic microcrystalline] [1 V dendritic]
	Nernst mobility equation	[branched $d_f = 1.76$ to $d_f = 2.0$]	[irregular microcrystalline]
high [3 V]	migration	[dense radial] dendritic	[3 V irregular microcrystalline]
rapid gas evolution [80–100 mA]	stirring	[stringy]	[porous irregular microcrystalline]

SCHEME 1

that may as the overpotential increases undergo reaction at the cathode (e.g. the evolution of hydrogen), one must consider general transport control by the Nernst mobility equation (Vetter 1967), the current density for a reacting species j, i_j,

$$i_j/nF = -D_j(\nabla C_j + (z_j C_j F/RT) \nabla \Phi) + C_j V, \qquad (2)$$

where Φ denotes the electric potential and the last term represents convection effects, which are absent in the present experiment. Equation (2) along with the electroneutrality condition needs be solved for all species

In the present experiment the minimum cell overpotential at which growth was observed was 0.4 V. The microstructure of these growths (Hibbert & Melrose 1988) clearly indicates activation control at this limit. The growths and paper have significant resistance (see below) and the overpotentials at the interface are reduced from those quoted (indeed the observed milliampere currents are small for the potentials applied). By theoretical estimates and direct observation we believe significant hydrogen evolution (more than 10%) is not present until cell overpotentials are in the range 0.8–1.0 V. There is, therefore, only a narrow window in the present experiment in which the diffusion limit can be achieved with indifferent support; however, significant concentration gradients are observed less than 0.8 V (see below). Fractals with fractal dimension approaching that of DLA were grown within this régime (Hibbert & Melrose 1988). Above 1.0 V, current against radius results are well described by an ohmic model (see below) that constitutes an approximation to (2). This implies migration control. Much smaller concentration gradients are present in this régime.

Although the zinc experiments showed a régime of centimetre-scale dendritic morphology, such a régime was not observed for copper in the paper support. In Hibbert & Melrose (1988), it was reported that high-voltage stringy growths were not observed. However, we have recently achieved this growth mode by passing relatively large (80 mA) currents through the cell and by allowing the dense radial growths to approach the anode; these will be reported on elsewhere.

OBSERVATION OF CONCENTRATION GRADIENTS

The paper support allows attempts to observe directly concentration gradients around the growths. We previously reported (Hibbert & Melrose 1988) the use of ammonia fumes in this context. Recently we have dropped the papers into NaOH solution quickly after growth; this precipitates out $Cu(OH)_2$. The turquoise blue precipitate usually forms exterior to the growth with a gap and slight gradation of colour around the growths. The gap region extends for some 2 mm/0.75 mm for 0.6 V/1.0 V respectively. These gradients are larger than the 0.02–0.05 mm thick diffusion layers observed in the absence of support. For 2 V and 3 V growths, no gaps were discernible to the eye, with the precipitate ending on the edge of the growth. The gaps observed by this method are somewhat less than those observed with the use of ammonia fumes (Hibbert & Melrose 1988).

Clearly at the low cell overpotentials (less than 0.8 V) significant concentration gradients do exist and this indicates some degree of diffusion control.

One 0.6 V growth investigated in this manner was highly asymmetric with growth considerably less developed on one side. This side was black in colour in contrast to the usual copper brown. On dropping the paper in NaOH the blue precipitate formed within the black side. Clearly this region was not electro-chemically active. The black colour is indicative of oxidation. This observation suggests a chemical effect to anisotropic growth at low voltage.

CURRENT AND DIAMETER AGAINST TIME: OHMIC MODEL

The current and growth diameter, D, were measured against time for growths at three different potentials. 1, 2 and 3 V. Figure 1 shows $\lg_2(D)$ and against $\lg_2(t)$ for these growths.

The diameter against time shows a power-law behaviour, $D \propto t^\beta$. The β exponents are 0.67, 0.62 and 0.54 respectively, these exponents are consistent with a $\beta = 1/d_f$ law. Such a law follows in the plane if one assumes that the depositional flux is proportional to the potential gradient (Matsushita *et al.* 1984). However, this implies, in the plane geometry, that the total current is constant. We have a contribution to the current from hydrogen evolution. We observe currents showing a positive curvature on logarithmic plots.

The variation of current in the 1–3 V overpotential régime fits a simple ohmic model. The growths and paper have significant resistance. Using a conductance bridge we measure for the solution-damp paper disc an anode-to-cathode

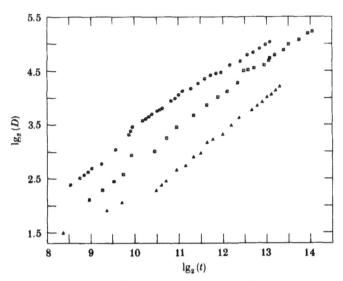

FIGURE 1. Plots of $\lg_2$(diameter) against $\lg_2$(time) (time in seconds) one volt triangles, two volt squares and three volt circles.

resistance *ca.* 500–700 Ω. A paper disc containing a growth has a lower resistance: the centre-to-perimeter resistance being some 50–70 % of a solution-damp control disc with the same radius. Note that the centre-to-perimeter resistance of a disc is logarithmically dependent on its radius and hence that the majority of its resistance is built up over the inner radii. The reference electrode is placed next to the anode. The potential drops across the paper and growth account for a significant portion of the fixed overpotential V. One finds that in the presence of a growth of radius L

$$V = V_g(L) + \eta(L) + V_p(L, L_a), \qquad (3)$$

where $V_g(L)$ is the potential drop across the disc of radius L of paper plus growth, $\eta(L)$ is the potential drop at the interface (including both activation and possibly concentration overpotential) and $V_p(L, L_a)$ is the drop across the solution damp paper annulus of inner radius L and outer radius equal to that of the anode L_a. One finds for a current $I(L)$ passing:

$$V_g(L) = I(L)\left((\rho_g/2\pi\sigma)\,Ln(L/L_0) + R_0\right) L > L_0, \qquad (4)$$

and

$$V_p(L) = I(L)\,(\rho_p/2\pi\sigma)\,Ln(L_a/L), \qquad (5)$$

where ρ_g and ρ_p are the resistivities of the growth plus paper and solution damp paper respectively and σ is the thickness of the disc. An inner cutoff has been applied to (4) with the resistance set at R_0 when the radius is L_0. Assume the existence of an L_0 such that we can approximate

$$\eta(L) \simeq \eta_c(\text{const.}) \quad \text{for} \quad L > L_0. \qquad (6)$$

Using (4) and (5) in (3) and eliminating the potential $V - \eta_c$ by taking the ratio $I(L_0)/I(L)$, one finds

$$I(L_0)/I(L) = A(1 - \rho_g/\rho_p) Ln(L_a/L) + B, \tag{7}$$

with A and B constants. Figure (2) shows a plot of $I(L_0)/I(L)$ against $Ln(L_a/2L)$ for the 2 V growth.

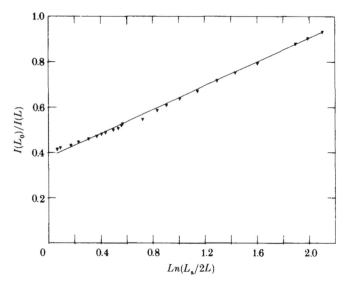

FIGURE 2. A plot of $I(L_0)/I(L)$ against $Ln(L_a/2L)$ for the 2 V growth.

The linearity of (7) is confirmed and was found to hold well for all three potentials investigated. L_0 in practice was chosen *ca*. 3 mm whereas the final growths were of radii *ca*. 15 mm. The constant A obeys $A > 0$ and the positive slope indicates that $\rho_g/\rho_p < 1$, as observed earlier.

The assumption (6) is justified *a posterior* and this suggests good control of the potential drop across the interface during all but the early stages of growth. However, this assumption needs be tested directly and we intend to run a cell with a roving electrode enabling direct measurements of the interface potential during growth.

A GROWTH INTERFACE

In an effort to probe the dependence of microstructure on growth conditions, a growth was performed at 3 V, the potential was switched off for 180 s then switched back on at 1 V and left to grow for several hours. The voltage was switched back up to 3 V for a short while before the growth was ended. Figure 3, plate 1, shows the resulting growth. Figure 4 shows the curve of measured current against time through the first change and for a while afterwards.

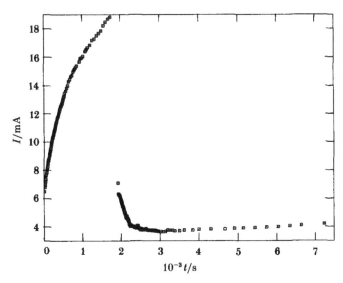

FIGURE 4. Current against time for the 3 V–1 V–3 V growth of figure 3.

After switching on the new voltage there is a steep drop in current over a period of 300 s, followed by a steady decline and levelling off. The current eventually starts to rise again after half an hour. If we assume the presence of an activation-limited process then a decline in the current can be explained as a decline in the active surface area as the active layer passes from the dense radial 3 V to the more open 1 V growth. Eventually the radial increase will compensate and the current will increase again.

The growth was subjected to analysis on a scanning electron microscope. Figure 5, plate 1, and figure 6, plate 2, show a typical growth from the 3 V region and a typical growth from the 1 V region respectively.

There is a clear morphological distinction. The 3 V microstructure of figure 5 has a branched structure (as seen under higher magnification) of irregularly arranged 0.5–1.0 µm sized grains and microcrystals. The 1 V microstructure of figure 6 has 1–5 µm sized plate and needle microcrystals arranged in a regular spine and side branch structure; we refer to this as dendritic microcrystalline (cf. Wranglen 1960). (We generally observe (110) dendrites, with the occasional (100). We have not observed more than the secondary branching evident in figure 6.)

A wedge-shaped portion of the growth was examined in the scanning electron microscope (SEM). By careful measurement the approximate location of the interface between the different growths was determined. Along three separate lines from the apex of the wedge we observed, at the approximate distance for the interface, a common structure unique to this region of the growth. Figure 7, plate 2, shows the best example. We are unable to observe similar effects at the second change of voltage.

Referring to figure 7, the irregular microcrystalline 3 V region on the left

3

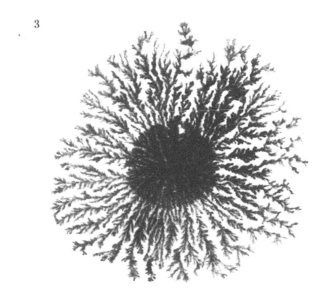

FIGURE 3. Photograph of the mixed potential growth. The average diameter of the growth is *ca.* 6 cm. Note the dense radial 3 V growth in the centre.

FIGURE 5. Electron micrograph of the microstructure in the 3 V region of the growth of figure 3. Scale bar represents 20 μm.

FIGURE 6. Electron micrograph of the microstructure in the 1 V region of the growth of figure 3. Scale bar represents 20 μm.

FIGURE 7. Electron micrograph of the structure at the interface of the 3 V and 1 V region of figure 3. Scale bar represents 50 μm.

terminates in a relatively compact mossy growth from which one observes a cascade of smaller mounds and in the extreme the emergence of a dendrite tip.

From the data of radius against time we estimate growth rates to be *ca.* 2.0–0.5 $\mu m\ s^{-1}$, the lower end of this range being appropriate for the 1 V growth at the diameter of the interface. It is reasonable to assume that the growth rate of the mossy growth is somewhat less than this. It is hence possible, given the scale of the mossy growth, that the steep decrease in current over 300 s corresponds to the growth of the mossy deposits.

The appearance of the mossy growth with a lag time before the appearance of the dendrite growth has been examined previously in the deposition of zinc (Diggle *et al.* 1969) and studied recently in computer simulations (Voss & Tomkiewicz 1985; Vicsek 1984) of activation-controlled growth. The regular structure of the dendrite tip being caused by lattice anistropy is of course not reflected in the simple simulations. The phenomena appearing here suggests that a degree of activation control of the Cu deposition is still present at the 1 V cell overpotential. Hydrogen evolution is also likely to be occurring.

A diffusion-based theory of this phenomenon has been developed (Bockris & Reddy 1973, and references therein). This is based on the decreasing radii of curvature of protruding tips. The theory is in essence a combination of the Mullins–Sekerka (1963) instability with a modelling of the dependence of activation overpotential on radii of curvature. The theory predicts a critical radius of curvature for dendritic growth. Figure 7 does demonstrate a successive decreasing radii of curvature of the cascading mounds. (There is no evidence in the micrograph for the mounds to ge generated by spiral and pyramid growth, as suggested by some theories (Diggle *et al.* 1969).)

Conclusions

In the experimental set up a region exists at low overpotential in which large concentration gradients can be observed, indicating diffusion control. However, because of the presence of hydrogen evolution, as the overpotential is increased the deposition becomes increasingly migration controlled. In the migration region an ohmic model correctly fits the current against radius characteristic. By growing successively at two different overpotentials we were able to generate a growth interface between irregular microcrystalline and dendritic microcrystalline deposits. The interface exhibited an unstable mossy deposit. This demonstrated the genesis of the branched growth, which can lead to fractal deposits on a larger scale.

J.R.M. would like to thank the Ministry of Defence for partial support, M. Perkins for permission to use facilities at Royal Holloway and Bedford New College, K. Singer for an idea, S. Janes and N. Paige for photographic assistance and D. Sherrington, S. Sarker and particularly Jenny for support and encouragement.

REFERENCES

Ball, R. C. 1986 In *On growth and form* (ed. H. E. Stanley & N. Ostrowsky), pp. 69–78. Dordrecht: Martinus-Nijhoff.

Ben-Jacob, E., Deutscher, G., Garik, P., Goldenfeld, N. D. & Lareah, Y. 1986 *Phys. Rev. Lett.* **57**, 1903–1906.

Bockris, J. O'M. & Reddy, A. K. N. 1973 *Modern electrochemistry*, vol. 2. New York: Plenum.

Bockris, J. O'M., Nagy, Z. & Drazic, D. 1973 *J. electrochem. Soc.* **120**, 30–41.

Brady, R. M. & Ball, R. C. 1984 *Nature, Lond.* **309**, 225–229.

Diggle, J. W., Despic, A. R. & Bockris, J. O'M. 1969 *J. electrochem. Soc.* **116**, 1503–1514.

Faust, J. W. 1968 *J. Cryst. Growth.* **34**, 433–435.

Giron, I. & Ogburn, F. 1961 *J. electrochem. Soc.* **108**, 842–847.

Grier, D., Ben-Jacob, E., Clarke, R. & Sander, L. M. 1986 *Phys. Rev. Lett.* **56**, 1264–1267.

Hibbert, D. B. & Melrose, J. R. 1988 *Phys. Rev.* A **38**, 1036.

Kahanda, G. L. M. K. S. & Tomkiewicz, M. 1988 *Phys. Rev.* B **38**, 957–959.

Kaufman, J. H., Nazzal, A. I., Melroy, O. & Kapitulnik, A. 1987 *Phys. Rev.* B **35**, 1881–1890.

Mandelbrot, B. B. 1982 *The fractal geometry of nature.* San Francisco: Freeman.

Mansfeld, F. & Gilman, S. 1970 *J. electrochem. Soc.* **117**, 1521–1523.

Matsushita, M., Hayakawa, Y. & Sawada, Y. 1985 *Phys. Rev.* A **32**, 3814–3816.

Matsushita, M., Sano, M., Hayakawa, Y., Honjo, H. & Swada, Y. 1984 *Phys. Rev. Lett.* **53**, 286–289.

Meakin, P. 1987 *Faraday Discuss. chem. Soc.* 1987 (Preprint.)

Mullins, W. W. & Sekerka, R. F. 1963 *J. appl. Phys.* **34**, 323–330.

Nittman, J. & Stanley, H. E. 1986 *Nature, Lond.* **321**, 663–668.

Swada, Y., Doughterty, A. & Gollub, J. P. 1986 *Phys. Rev. Lett.* **56**, 1260–1263.

Vetter, K. J. 1967 *Electrochemical kinetics.* New York: Academic Press.

Vicsek, T. 1984 *Phys. Rev. Lett.* **53**, 2281–2284.

Voss, R. F. & Tomkiewicz, M. 1985 *J. electrochem. Soc.* **132**, 371–375.

Wranglen, G. 1960 *Electrochim. Acta* **2**, 130–143.

Witten, T. A. & Sander, L. M. 1983 *Phys. Rev.* B **27**, 5686–5697.

Discussion

R. C. BALL (*Cavendish Laboratory, University of Cambridge, U.K.*). Does Dr Melrose see any evidence of restructuring after growth?

J. R. MELROSE. Restructuring of electrodeposits left in the experimental solution can be expected as a result of concentration-cell effects Wranglen (1960). The deposits in paper reported above were washed and dried at room temperature a short time (less than 30 s) after the potential was switched off. The electron microscope investigations took place from one to three weeks after growth. I have not seen any evidence I could interpret as restructuring after growth. A control experiment leaving the growths in the experimental solution for a range of time periods was not conducted.

Flow through porous media: limits of fractal patterns

By R. Lenormand†

Dowell Schlumberger, B.P. 90, 42003 Saint Etienne Cedex 1, France

By using experiments on micromodels and computer simulations, we have demonstrated the existence of three types of basic displacements when a non-wetting fluid invades a two-dimensional porous medium. capillary fingering when capillary forces are very strong compared to viscous forces, viscous fingering when a less viscous fluid is displacing a more viscous one, and stable displacement in the opposite case. These displacements are described by statistical models: invasion percolation, diffusion-limited aggregation (DLA) and anti-DLA.

The domains of validity of the basic displacements are mapped onto the plane with axes Ca (capillary number) and M (viscosity ratio). The boundaries of these domains are calculated either by using theoretical laws describing transport properties of fractal patterns or by the interpretation of physical mechanisms at the pore scale. In addition, the prefactors that are not available from scaling theories are obtained by computer simulations on a network of capillaries, in which the flow equations are solved at each node.

1. INTRODUCTION

The purpose of this paper is to provide better understanding of the relevant mechanisms that control the displacement of a wetting fluid by a non-wetting fluid in a porous medium when both capillary and viscous forces are present.

By using experiments on micromodels and computer simulations, we have previously demonstrated the existence of three types of basic displacements:

(*a*) capillary fingering when capillary forces are very strong compared to viscous forces;

(*b*) viscous fingering when a less viscous fluid is displacing a more viscous one;

(*c*) stable displacement in the opposite case.

We have also shown how these displacements can be described by statistical models: invasion percolation, diffusion-limited aggregation (DLA) and anti-DLA (Lenormand *et al.* 1988). These types of displacements have also been observed and studied separately by other authors (see, for example, Leclerc & Neale 1988, for a detailed bibliography), but the novel feature of our approach is to include all these various mechanisms in a general model.

The principle of our approach is to consider these displacements as three limiting cases of more general displacements involving the simultaneous competition between viscous forces in injected and displaced fluid, and capillary forces. The relative intensity of these three types of forces is characterized by two dimensionless numbers, the ratio M of the viscosities of the two fluids ($M = \mu_2/\mu_1$,

† Present address Institut Français du Pétrole. 1 & 4 av. de Bois-Préau. B P 311, 92506 Rueil Maimaison Cedex. France

the fluid 2 being the injected fluid), and the capillary number Ca, which is the ratio of viscous forces to capillary forces:

$$Ca = \mu_2 q / \Sigma \gamma, \tag{1}$$

q being the flow rate through an area Σ of the medium and γ the interfacial tension.

The domains of validity of the basic displacements are mapped onto the plane with axes Ca and M, and this mapping represents the phase diagram for two-dimensional displacements in drainage condition (figure 1).

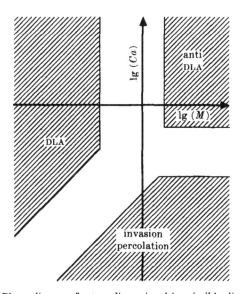

FIGURE 1. Phase diagram for two-dimensional immiscible displacements.

The main purpose of this paper is to calculate the boundaries of the domains of the phase diagram (figure 1). The influence of the various parameters (especially the size of the network) is obtained either by using theoretical laws describing transport properties of fractal patterns or by the interpretation of physical mechanisms at the pore scale. In addition, the prefactors that are not available from scaling theories are obtained by computer simulations on a network of capillaries, in which the flow equations are solved at each node.

2. BASIC IMMISCIBLE DISPLACEMEMENTS

In this section, we recall the results of micromodel experiments and computer simulations obtained for limiting values of the parameters M and Ca (Lenormand et al. 1988). The corresponding statistical models, which will be used in the next section for the calculation of the theoetical diagram, are also briefly presented.

2.1. *Capillary fingering and invasion percolation*

This approach is related to capillary mechanisms that take place at the micro-scopic (pore) scale, when viscous forces are negligible (quasi-static displacement).

In drainage régime, capillary forces prevent the non-wetting fluid from spontaneously entering a porous medium. For a given difference of pressure P_0 between the fluids, the injected fluid can only enter the throats larger than R_0 given by Laplace's law (with perfect wetting, $\cos\theta = 1$):

$$P_0 = 2\gamma/R_0. \tag{2}$$

Assuming that the porous medium can be described by a two-dimensional network of pores connected by throats (sites and bonds), then, from a statistical point of view, a throat of size $R > R_0$ is a conductive bond and a throat with $R < R_0$ a non-conductive bond. The fraction p of active bonds can easily be deduced from the throat size distribution. At a given pressure P_c, which corresponds to the critical fraction p_c, the injected fluid invades all the channels connected to the injection face; this mechanism is called invasion percolation and has been described by several authors (Wilkinson & Willemsen 1983; Chayes *et al.* 1985; Lenormand & Zarcone 1985).

Figure 2 shows a comparison between an experiment on a micromodel (*a*), a computer simulation (*b*) and the pattern produced with the model of invasion percolation (*c*).

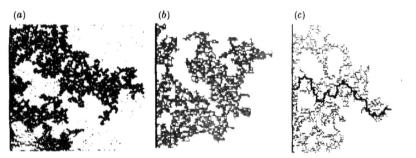

FIGURE 2. Capillary fingering: (*a*) experiment, (*b*) network simulation, and (*c*) statistical model (invasion percolation).

Invasion percolation is a local model, i.e. at each step, the injected fluid invades one pore (or site) according to the size of the throats along the interface between the fluids. At a given stage, menisci in the pores do not see the exit because we are assuming a zero pressure drop in the fluids.

The physics is different for viscous displacements.

2.2. *Viscous displacements*: DLA *and anti-*DLA

Viscous displacements, either stable or unstable, are governed by the pressure field between the entrance and the exit. Consequently, even for a stable displacement, a local model based on some rules at the interface cannot be used

for modelling viscous displacements. A model, called gradient-governed growth
(GGG) has been developed simultaneously by several authors (DeGregoria 1985;
Sherwood & Nittmann 1986) to solve this problem, by using both a continuum
approach to calculate the pressure field and a discrete displacement of the
interface which accounts for the granular structure of the porous medium. Let us
examine the two limiting cases when $M \to \infty$ and $M \to 0$.

Unstable viscous displacement $(M \to 0)$: DLA

In this limit of zero viscosity ratio the pattern obtained during a displacement
can be represented by a model known as diffusion-limited aggregation (DLA)
(Paterson 1984; Witten & Sander 1983; Måløy *et al.* 1985). A computation based
on a network of random conductances, leads to similar results (Chen *et al.* 1985).

A comparison between an experiment, a network simulation and a pattern
produced by DLA is shown in figure 3.

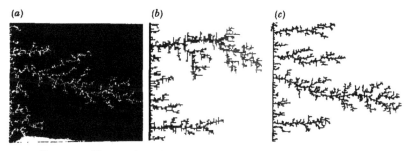

FIGURE 3. Viscous fingering: (*a*) experiment, (*b*) network simulation,
and (*c*) statistical model (DLA).

Stable viscous displacement $(M \to \infty)$: anti-DLA

There is no fingering and trapping of the wetting phase is very low. In this case
the GGG model becomes identical to anti-DLA (Paterson 1984). It consists in
releasing anti-particles (equivalent to the injected fluid) on a face of the network
where all the sites are occupied by particles. The anti-particle moves at random

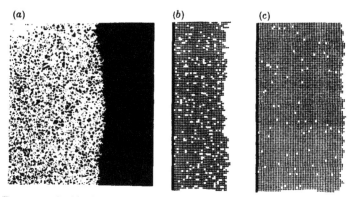

FIGURE 4. Stable displacement: (*a*) experiment, (*b*) network simulation,
and (*c*) statistical model (anti-DLA).

on the lattice until it reaches an occupied site. Then, both the particle and the anti-particle are removed and another anti-particle is launched at the entrance. Figure 4c shows the results of an anti-DLA simulation on a 100 × 100 network, where we added a condition for taking into account trapping of the displaced fluid. Figure 4a, b shows experimental and computed results under the same flow conditions.

3. BOUNDARIES OF THE DOMAINS

These three basic displacements can be understood as theoretical limit when two of the three kinds of forces involved during the displacement are negligible. Theoretically these limits cannot be reached in an infinite medium. However, as shown in the two-dimensional experiments, they can be observed in a finite system, inside a domain of the plane Ca–M. The purpose of the following section is to calculate the frontiers of these domains in order to answer the important question: for given conditions of injection, which kind of displacement should be expected?

A simplified version of this calculation has previously been presented, but without taking into account the fractal nature of the patterns in percolation and DLA domains (Lenormand 1985, 1986).

3.1. *Principle of the calculation*

As previously described, each basic displacement is characteristic of a dominant type of force: viscous forces in the displacing fluid for stable displacement, viscous forces in the displaced fluid for viscous fingering and capillary forces for capillary fingering. Each domain will be limited by two frontiers in the plane Ca–M, each of them corresponding to the perturbation induced by one of the two other kinds of force.

Let us call S (macroscopic saturation) the fraction $N_b/2L^2$ of invaded bonds at breakthrough. By using the network simulator (Lenormand *et al.* 1988), we have shown that the saturation S evolves between two plateaux when M or Ca is varied (figure 5). Each plateau corresponds to one of the three basic domains. We define the limit M^* (or Ca^*) of each domain such as

$$|S(M^*) - S(\text{plateau})| = \epsilon, \tag{3}$$

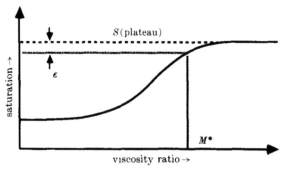

FIGURE 5. Fraction of invading fluid (saturation) at breakthrough as a function of M (100 × 100 network simulation)

ϵ being a small constant, with the same value for all the boundaries (say for instance $\epsilon = \frac{1}{100}$).

3.2. Geometrical properties of the medium

The porous medium is described by a two-dimensional network of inter-connected capillaries with radius R varying uniformly in the interval $[(1-\sigma)R_0, (1+\sigma)R_0]$. The macroscopic size of the network is $L \times L$ and the mesh size is a (microscopic length scale).

For a monophasic flow, the pressure gradient $\Delta P/x$ across the network is linked to the volumetric flow rate q, the permeability k, the cross-sectional area of the flow Σ and the viscosity μ of the fluid by Darcy's law

$$\Delta P/x = \mu q/k\Sigma. \tag{4}$$

Resistance to flow can also be calculated as a function of the mean conductance g_0 of a single channel (Poiseuille's law)

$$g_0 = \tfrac{1}{8}\pi(R_0^4/a\mu) \tag{5}$$

and, by comparison of this equation with Darcy's law calculated at pore level $(\Sigma = a^2; x = a)$

$$k = \tfrac{1}{8}\pi(R_0^4/a^2). \tag{6}$$

3.3. Boundaries of the percolation domain

Let us consider the quasi-static limit of the displacement $(Ca \to 0)$, as it can be simulated with a computer. N_b is the number of displaced channels at breakthrough.

3.3.1. Injecting the more viscous fluid

Close to the percolation threshold, the injected fluid flows through the backbone of the percolation cluster, and, because of viscous effects, the pressure is larger near the entrance. Consequently, some channels, with a size smaller than the threshold value R_c, can be now invaded and the number of invaded channels increases by a value ΔN. The pattern is more compact near the entrance in comparison with the pure percolation case. Increasing the flow rate will progressively lead toward a stable displacement. The problem is now to calculate ΔN as a function of the flow conditions.

The following calculation is based on the classical properties of percolation clusters (Wilkinson et al. 1983). The percolation threshold for a network of size L is p_c, larger than the theoretical value $p^* = 0.5$ for an infinite square network. Transport and geometrical properties of the percolation cluster are linked by the following relations:

(a) correlation length

$$L/a \propto (p_c - p^*)^{-\nu}, \tag{7}$$

(b) size of the percolation cluster

$$N_b/L^2 \propto (p_c - p^*)^\beta, \tag{8}$$

conductivity $\qquad\qquad G/g_0 \propto (p_c - p^*)^t. \tag{9}$

Now, for a given rate of injection q, viscous effects lead to an increase of pressure at the entrance

$$\Delta P = q/G. \tag{10}$$

By using Laplace's law (equation (2)), the corresponding variation of size of accessible throats is

$$\Delta R/R_0 = R_0 \Delta P/\gamma, \tag{11}$$

which leads to a variation Δp of the fraction of accessible bonds (the size distribution is uniform over a width $2\sigma R_0$)

$$\Delta p = \Delta R/2\sigma R_0. \tag{12}$$

The corresponding variation of bonds in the cluster is obtained by derivation of equation (8)

$$\Delta N/L^2 \propto (p_c - p^*)^{\beta-1} \Delta p, \tag{13}$$

and, by combination of the previous equations

$$\frac{\Delta N}{L^2} \propto \left(\frac{L}{a}\right)^{(1+t-\beta)/\nu} \frac{a^3}{R_0^3} \frac{L}{a} \frac{1}{\sigma} \frac{\mu_2 q}{aL\gamma}. \tag{14}$$

The last term of this equation is by definition the capillary number Ca (equation (1), with a cross-sectional area equal to aL). To obtain its value Ca^* on the boundary, we write that the relative variation $\Delta N/N_b$ of the number of bonds in the percolation cluster is equal to ϵ

$$\left(\frac{L}{a}\right)^{(1+t+\nu)/\nu} Ca^* \frac{a^3}{R_0^3} \frac{1}{\sigma} = \epsilon, \tag{15}$$

or, finally

$$Ca^* = A\epsilon\sigma \left(\frac{L}{a}\right)^\tau \left(\frac{R_0}{a}\right)^3, \tag{16}$$

where $\tau = -(t+1+\nu)/\nu$. Accepted values for two-dimensional percolation are $t \approx \nu = \frac{4}{3}$, and therefore $\tau = -2.75$. The prefactor A can be estimated from the numerical simulations obtained for a square network (Lenormand *et al.* 1988). For the 100×100 network, $Ca^* \approx 10^{-6}$, $R_0 = 0.23$ mm, $\sigma = 0.56$, $a = 1$ mm and $A \approx 5000$.

3.3.2. *Injecting the less viscous fluid*

The capillary pressure is decreased by ΔP because of the pressure drop in the displaced fluid. Consequently, the fraction of accessible channels decreases by a value Δp and the fingers become thinner and thinner when the flow rate increases (towards the DLA domain).

The pressure drop is calculated by assuming that the displaced fluid can flow in the whole network, without reduction of permeability due to the branches formed by the injected fluid. From equation (4)

$$\Delta P = \mu_1 qL/kaL. \tag{17}$$

Now, using the same method as previously, with $\mu_1 = \mu_2/M$, we obtain the equation of the boundary of percolation towards DLA

$$Ca^* = Be\sigma \left(\frac{L}{a}\right)^{-(\nu+1)/\nu} \left(\frac{R_0}{a}\right)^3 M. \tag{18}$$

The value of the prefactor is $B \approx 800$, estimated from the 100×100 simulation.

3.4. Boundaries of the DLA domain

The analogy between viscous fingering and DLA simulations requires the following three conditions.

(1) No pressure drop in the injected fluid (limit $M \to 0$).

(2) No pressure jump at the interface between the fluids.

(3) Large randomness due to the pore size distribution.

Conditions (1) and (2) determine the two frontiers of the DLA domain, the viscous and the capillary limits. Condition (3) is assumed to be always satisfied in a porous medium.

3.4.1. Viscous limit

The injected fluid flows both at the tip of the tree-like finger and everywhere on the interface between the fluids with respective flow rates q' and q''. In a pure DLA régime only the tip of the finger is growing (screening effect) and consequently $q'' = 0$. We can estimate the flow rates by using the approximations of two independent flows, one in the finger (replaced by a straight channel, section a^2) and the other in the whole network (section aL). The equality of the pressures (equation (4)) in each fluid at any distance from the entrance leads to

$$q''/q' = ML/a \tag{19}$$

During an injection time t, the number of invaded bonds N_b is equal to $q't/v_0$, where v_0 is the mean volume of a pore. The excess of bonds ΔN due to viscous forces in the injected fluid is $q''t/v_0$. Consequently: $\epsilon = q''/q'$ and

$$M^* = \epsilon a/L. \tag{20}$$

This result is in good agreement with the numerical simulations.

3.4.2. Capillary limit

Condition (2) requires no pressure jump at the interface between the fluids. More accurately, a constant and small pressure jump would not perturb a DLA-type displacement (the interface is always an equipotential), but fluctuations from pore to pore must be negligible in comparison with the viscous pressure drop in the displaced fluid at pore level

$$\gamma \sigma R_0/R_0^2 \ll \mu_1 qa/ka^2. \tag{21}$$

This leads to the limiting value

$$Ca^* = Ce\sigma(a/L)(R_0/a)^3 M, \tag{22}$$

where the numerical value for C can be estimated to be 1300.

3.5. *Boundaries of the anti-*DLA *domain*

In this domain, the dominant force is related to the viscosity of the injected fluid and perturbations are provided by capillary effects or viscous effects in the displaced fluid.

3.5.1. *Capillary limit*

The injected fluid invades all the channels in the network. Therefore, capillary effects are caused by the difference of capillary pressure between the smallest and the largest channels. As before, we assume that the variation of capillary pressure is small compared with the viscous pressure drop at pore scale. However, contrary to the fingering case, we observe a displacement in all the pores on the frontier at the same time

$$\gamma \sigma R_0 / R_0^2 \ll \mu_2 qa/kaL. \tag{23}$$

Therefore, Ca^* is given by the following equation where the prefactor is estimated as before:

$$Ca^* = 80\epsilon\sigma(R_0/a)^3. \tag{24}$$

3.5.2. *Viscous limit*

In this limit, capillary forces are assumed to be zero. However, this case is not a miscible displacement and the fluids cannot mix. Consequently, there is no characteristic timescale linked to molecular diffusion and, for a creeping flow (no inertial forces), the displacement pattern is independent of the flow rate. Therefore, the frontier is parallel to the Ca axis.

Comparison of viscous forces in both fluids at pore scale leads to the condition $\mu_2 > \mu_1/\epsilon$ or

$$M^* \approx 1000\epsilon. \tag{25}$$

Further simulations with a GGG model must be performed for a better understanding of this viscous limit and especially to describe the effect of pore-size distribution.

4. CONCLUSION

We have presented a theoretical description of immiscible displacements (drainage) in two-dimensional porous media. We have shown how the recent stochastic models of percolation, diffusion limited aggregation and anti-DLA can be included in a general description of displacement patterns.

The main result is the calculation of a diagram that displays the boundaries of three domains corresponding to the basic mechanisms: percolation, DLA and anti-DLA. The limits of these domains have been calculated by taking into account the fractal nature of the patterns and the physical mechanisms governing the displacements.

Concerning the scaling of the diagram with the size L of a two-dimensional network, the main conclusions are the following.

(1) Stable displacement boundaries do not depend on the size of the network.

(2) Percolation boundaries depend strongly on the size ($L^{-1.75}$ and $L^{-2.75}$). Consequently, this régime should be difficult to observe at large scale.

(3) DLA boundaries depend weakly on the size (L^{-1}). Moreover, this domain extends toward percolation when the size increases. This kind of displacement should be easily observed at large scale.

REFERENCES

Chayes, J. T., Chayes, L. & Newman, C. M. 1985 *Communs math. Phys.* **101**, 383–407.
Chen, J. D. & Wilkinson, D. 1985 *Phys. Rev. Lett.* **55**, 1892–1895.
DeGregoria, A. J. 1985 *Physics Fluids* **28**, 2933–2935.
Leclerc, D. F. & Neale, G. H. 1988 *J. Phys.* A **21**, 2979–2994.
Lenormand, R. 1985 *C.r. acad. Sci., Paris* II **301**, 247–250.
Lenormand, R. 1986 In *61st Annual Tech. Conf. and Exhib. of SPE, New Orleans.*
Lenormand, R., Touboul, E. & Zarcone, C. 1988 *J. Fluid Mech.* **189**, 165–187.
Lenormand, R. & Zarcone, C. 1985 *Phys. Rev. Lett.* **54**, 2226–2229.
Måløy, K. J., Feder, J. & Jøssang, T. 1985 *Phys. Rev. Lett.* **55**, 1885–1891.
Paterson, L. 1984 *Phys. Rev. Lett.* **52**, 1621–1624.
Sherwood, J. D. & Nittmann, J. 1986 *J. Phys., Paris* **47**, 15–22.
Wilkinson, D. & Willemsen, J. F. 1983 *J. Phys.* A **16**, 3365–3376.
Witten, T. A. & Sander, L. M. 1983 *Phys. Rev.* B **27**, 5686–5697.

Fractal BET and FHH theories of adsorption: a comparative study

By P. Pfeifer,[1] M. Obert[1]† and M. W. Cole[2]

[1] Department of Physics, University of Missouri, Columbia,
Missouri 65211, U.S.A.
[2] Department of Physics, The Pennsylvania State University,
University Park, Pennsylvania 16802, U.S.A.

Two theories of multilayer adsorption of gases, namely the Brunauer–Emmett–Teller (BET) theory and the Frenkel–Halsey–Hill (FHH) theory, have recently been extended to the case of fractal substrates in a number of different ways. We present a critical evaluation of the various predictions. The principal results are the following. At high coverage, the fractal BET and FHH isotherms apply to mass and surface fractals, respectively. Both give characteristic power laws with D-dependent exponents (D = fractal dimension of the substrate). The BET isotherm additionally depends on the topological dimension D_{top} of the substrate. For fractal aggregates ($D_{top} = 1$) with $D < 2$, the adsorbed phase exists only in a highly disordered state. The BET theory is sensitive to multiple-wall effects (they affect prefactors); the FHH theory is not. For the FHH theory, detailed assessments of the approximations in the model are available. The predictions of the FHH theory have been observed on fractal silver surfaces.

1. Introduction

Since the first exploration of fractal surface properties of solids at molecular scales (Avnir & Pfeifer 1983) experimental investigations have uncovered a wealth of materials with a well-defined fractal surface dimension D (see, for example Avnir *et al.* 1984, 1985; Schaefer *et al.* 1987; Pfeifer 1987; Schmidt 1988; Farin & Avnir 1989), as determined by a wide variety of techniques. Properties that are sensitive to D include small-angle X-ray and neutron scattering (Bale & Schmidt 1984; Wong *et al.* 1986; Martin & Hurd 1987), multiple scattering and absorption of light (Berry & Percival 1986), electronic energy transfer between adsorbed molecules (Klafter & Blumen 1984; Even *et al.* 1984; Pines *et al.* 1988), pore-size distribution (Pfeifer & Avnir 1983; Pfeifer 1987; Frank *et al.* 1987; Spindler *et al.* 1987), diffusion in pore space (de Gennes 1982; Pfeifer *et al.* 1984; Nyikos & Pajkossi 1986), alternating current (AC) response of electrodes (Liu 1985; Nyikos & Pajkossi 1985; Vlachopoulos *et al.* 1988), and Coulomb screening in ionic layers (Blender & Dieterich 1986).

In this paper we describe recent progress in the understanding of thermodynamic properties of adsorbed films on fractal surfaces. The situation to be addressed is the one where the surface is fractal from a few ångströms to scales of

† Present address: Department of Chemistry, Universität Giessen, D-6300 Giessen, FRG

say 100 Å,† and where the adsorbed film varies in thickness over the same range of scales and is in equilibrium with its own vapour. The motivation for this study is the following.

1. Adsorbed films on irregular surfaces in this range form the basis of many important processes in science and technology. Condensation phenomena, wetting and most of surface chemistry, including catalysis and corrosion, fall into this category. Therefore knowledge of the thermodynamic behaviour of the adsorbed phase is fundamental for the understanding of any of these properties. We will be concerned with adsorption isotherms. They give complete knowledge of the configurational partition function of the adsorbed phase (Steele 1974).

2. The availability of fractal materials in this range (or alternatively, the availability of methods to examine a given material for fractality in this range) makes it possible to compare the predictions with experiment. Such tests are of interest because none of the traditional theories of multilayer adsorption, which are:

(i) the Brunauer–Emmett–Teller theory (BET, planar surface, film thickness less than about 1.5 layers);

(ii) the Frenkel–Halsey–Hill theory (FHH, planar surface, film thickness more than about 1.5 layers);

(iii) capillary condensation (enclosed space, pore of diameter more than about 50 Å, film thickness of order of pore diameter);

is directly available. Thus there is little guidance from established theories as to what to expect on non-planar substrates. A successful description of multilayer adsorption on fractal substrates will open up a new régime of structural information available from isotherm measurements.

3. In many earlier studies of fractal materials (Avnir et al. 1984, 1985) the fractal nature of the surface was established by demonstrating that over some range of diameters, a_{min} to a_{max}, the number of molecules of diameter a required to cover the surface with a monolayer, N_m, satisfies

$$N_m = Ca^{-D} \tag{1}$$

(or some equivalent thereof). In (1) D is the fractal dimension of the surface, $2 \leqslant D \leqslant 3$, and C measures the D-dimensional content (Hausdorff measure) of the surface. For a planar surface, $D = 2$ and C is approximately the surface area. The monolayer values N_m in those studies were obtained from isotherm analyses such as (i), the argument being that the functional form of the isotherm at low coverage is insensitive to surface irregularity. This argument has recently been questioned (Fripiat et al. 1986). Here we show that such doubts are unfounded.

The starting point of our investigation are the fractal BET theories by Cole et al. (1986) and by Fripiat et al. (1986), and two fractal FHH theories by Pfeifer et al. (1989). We show that these generalizations of the classical BET and FHH isotherms apply to complementary situations, namely to mass and surface fractals, respectively (table 1). They reflect the fact that very different surface–adsorbate

† 1 Å = 10^{-1} nm = 10^{-10} m.

TABLE 1. DOMAINS OF APPLICABILITY AND OTHER FEATURES OF THE FRACTAL
BET AND FHH ISOTHERMS

(The solid is a mass fractal when both the surface and the entire solid scale like a D-dimensional system. The solid is a surface fractal when the surface D-dimensional and the solid is three-dimensional Low and high coverage refers to film thicknesses of less than about 1 5 layers and more than about 1.5 layers, respectively.)

	BET isotherm	FHH isotherm
surface–adsorbate potential	short-range, controls first layer only	long range, controls all layers
source of potential	surface atoms	all atoms in solid (semi-infinite solid)
at low coverage, applies to	mass/surface fractals	—
at high coverage, applies to	mass fractals	surface fractals
multilayer formation driven by	entropy	energy
film–vapour interface	rough	smooth
associated lattice–gas model	exactly solvable on planar surface	approximate

interactions are dominant depending on whether the solid is a mass or surface fractal. We show that the two isotherms follow an accordingly different asymptotic behaviour at high coverage. Motivated by the recent discovery of solids which are mass fractals with $D < 2$ in the range of scales at issue (Schaefer *et al.* 1987), we extend the BET theory to include the case $D < 2$ and show that the thermodynamic behaviour is very different from the $D > 2$ case. For the fractal FHH isotherm, we show that surface-tension effects can safely be neglected up to a film thickness of 50 Å or more, depending on the substrate. With regard to the two competing versions of the BET isotherm, we show that they are identical at very low coverage, but differ by as much as a factor of two at high coverage. The two versions of the FHH isotherm, in contrast, are strictly proportional to each other at all coverages.

2. BET THEORY ON FRACTAL SURFACES

In the BET model for a planar surface (see, for example, Steele 1974) the energy of an adsorbate configuration with N_1 molecules in the first layer and N_2 molecules in higher layers is given by $N_1 \epsilon_1 + N_2 \epsilon_2$ where ϵ_1 and ϵ_2 are the heat of adsorption and the heat of condensation (bulk) per molecule. This shows that in a lattice representation the model is also well-defined for any non-planar surface, and that in the model the surface–adsorbate potential extends to the first layer only (ϵ_1), all higher layers being governed by adsorbate–adsorbate interactions instead (ϵ_2). Thus, the model is naturally restricted to situations where long-range van der Waals interactions between the solid and the absorbate are unimportant, which is at low coverage ($N_1 > N_2$) or if the solid consists mostly of surface, assumed to be energetically uniform. In the latter case ($N_2 \gg N_1$), the solid–adsorbate interactions are outnumbered by the adsorbate–adsorbate interactions. The model leads to a rough film–vapour interface, even on a planar surface, because the only way higher layers can be more stable than bulk liquid is by having a higher entropy (table 1). This produces fluctuations in film thickness (coexistence of patches of different thickness). The model for a planar surface ($D = 2$) predicts that the number of adsorbed molecules, N, grows with increasing gas pressure p as

$$N = N_m \frac{cx}{1-x} \frac{1}{1+(c-1)x},$$ (2)

$$c \equiv \exp[-(\epsilon_1 - \epsilon_2)/kT] \quad (c > 1),$$ (3a)

$$x \equiv p/p_0 \quad\quad\quad (0 \leqslant x \leqslant 1),$$ (3b)

where N_m is the number of surface sites, k is Boltzmann's constant, T is the temperature and p_0 is the saturation pressure (coexistence pressure for vapour and bulk liquid). Thus N/N_m gives the mean film thickness (in units of the adsorbate diameter a), c is the adsorption strength, and $x \rightarrow 1$ corresponds to the condensation of bulk liquid on the surface. The success of this classical BET isotherm, (2), in describing low-pressure experimental isotherms ($x < 0.3$) for an enormous variety of systems has made it the most widely used isotherm in surface science (determination of N_m). This success makes it natural to ask what the BET model predicts for a fractal surface, i.e. for $D > 2$.

Two rather distinct approaches to this question have been developed. One is based on the idea that as the film grows in thickness, it fills pores (irregularities) and offers fewer sites for further adsorption because the film–vapour interface shrinks. By using this idea, Fripiat et al. obtained

$$N/N_m = \frac{c}{1+(c-1)x} \sum_{n=1}^{\infty} n^{2-D} x^n$$ (4)

for the fractal generalization of the BET isotherm (Fripiat et al. 1986; Levitz et al. 1988; Fripiat 1989). We give a simplified version of their calculation and show that (4) misses two important points. Following the usual derivation of (2), one divides the adsorbed phase into patches consisting of $n = 1, 2, \ldots$ layers, respectively. The n-layer patch by definition consists of all film segments for which the film–vapour interface has distance na from the surface. Denote the number of molecules in the bottom and top layer of the n-layer patch by S_n and S'_n, respectively. For a planar surface one has $S'_n = S_n$, but for an irregular surface $S'_n < S_n$ ($n \geqslant 2$) as a result of pore filling. The relation between S'_n and S_n for a fractal surface is

$$S'_n = n^{2-D} S_n \quad (n = 1, 2, \ldots).$$ (5)

This follows from a simple application of (1) (S'_n is proportional to the number of particles of size na needed to cover the surface, times the cross section area of a particle). From (5), one obtains the total number of molecules in the film as

$$N = \sum_{n=1}^{\infty} (S_n + 2^{2-D} S_n + \ldots + n^{2-D} S_n)$$ (6a)

$$= \sum_{n=1}^{\infty} n^{2-D}(S_n + S_{n+1} + \ldots).$$ (6b)

In (6a), the summation over n runs over all patches, and the expression $(\ldots)$

counts the molecules in a patch layer by layer. Analogously to (6a), the total number of adsorption sites on the solid satisfies

$$N_m = \sum_{n=0}^{\infty} S_n, \tag{7}$$

where S_0 is the number of bare surface sites. Notice that for a finite system, (5) holds only up to $n \sim R_{max}/a$ where R_{max} is the outer cutoff of the fractal régime (radius of largest pore). Thus in (6) and (7) we have idealized the system as infinite, $R_{max} \to \infty$ (thermodynamic limit). The equilibrium condition is that the rate of condensation into the nth layer of the film must equal the rate of evaporation from the nth layer ($n = 1, 2, \ldots$). On a planar surface this amounts to

$$cxS_0 = S_1 \quad (n = 1), \tag{8a}$$

$$xS_{n-1} = S_n \quad (n = 2, 3, \ldots). \tag{8b}$$

Equations (8) state that the condensation rate is proportional to the vapour pressure and to the number of exposed molecules in the $(n-1)$th layer, that the evaporation rate is proportional to the number of exposed molecules in the nth layer, and that for $n = 1$ the condensation rate is also proportional to the adsorption strength. Equations (8) yield

$$S_n = cx^n S_0 \quad (n = 1, 2, \ldots). \tag{9}$$

The assumption made by Fripiat *et al.* is that (8) and (9) also hold on a fractal surface. Substitutions of (9) into (6b) and (7) gives their result, equation (4).

The problem is, however, that on a fractal surface the number of exposed molecules in layer n is given by S'_n, not by S_n. Therefore (8b) should be replaced by $xS'_{n-1} = S'_n$ and (9) by $S'_n = cx^n S_0$ (note $S'_1 = S_1$). Together with (5) this gives

$$S_n = cx^n n^{D-2} S_0 \quad (n = 1, 2, \ldots) \tag{9'}$$

in place of (9). From (9'), (6b) and (7) we obtain

$$N/N_m = \left(c \sum_{n=1}^{\infty} n^{2-D} \sum_{j=n}^{\infty} j^{D-2} x^j \right) \bigg/ \left(1 + c \sum_{j=1}^{\infty} j^{D-2} x^j \right) \tag{4'}$$

for a corrected version of (4). The inequality

$$\frac{x^n}{1-x} \leqslant \sum_{j=n}^{\infty} j^{D-2} x^j \leqslant x \frac{d}{dx} \frac{x^n}{1-x} \quad (2 \leqslant D \leqslant 3) \tag{10}$$

shows that the series in (4') converge for all $0 \leqslant x < 1$. The second deficiency of the isotherm (4) is that it neglects 'multiple-wall' effects. It treats the filling of pores without taking into account that, as a pore is being filled, the film grows from two opposite walls and stops growing when the two films meet (at the latest). Indeed, in the theory presented so far the surface geometry enters only through (5) (decrease of film area with increasing film thickness), which is sensitive to multiple walls only in an indirect manner. It follows that the isotherm (4') has the same deficiency.

We therefore turn to the second approach to the fractal BET problem (Cole *et al.* 1986) which explicitly includes multiple-wall effects. It replaces the D-dimensional surface by an equivalent system of pores, each of which can be treated as independent and in each of which the coverage can be calculated from the BET theory for adsorption between two plates (maximum of n layers where $2na$ is approximately the pore diameter). The total number of adsorbed molecules is obtained by integrating the contributions from the individual pores. Thus D here enters through the pore-size distribution of a fractal surface. The result in the thermodynamic limit is

$$N/N_{\mathrm{m}} = (D-2)\int_1^\infty n^{1-D} f_n(x)\,\mathrm{d}n, \tag{11}$$

where
$$f_n(x) \equiv \left(c\sum_{j-1}^n jx^j\right)\Big/\left(1+c\sum_{j-1}^n x^j\right) \qquad (n = 1, 2, \ldots) \tag{12a}$$

$$= \frac{cx}{1-x}\frac{1-(n+1)x^n+nx^{n+1}}{1+(c-1)x-cx^{n+1}} \qquad (1 \leqslant n \leqslant \infty) \tag{12b}$$

$$= x\frac{\mathrm{d}}{\mathrm{d}x}\ln\left(1+cx(1-x^n)/(1-x)\right) \qquad (1 \leqslant n \leqslant \infty). \tag{12c}$$

The function $f_n(x)$ is the planar BET isotherm for a maximum of n layers (cf. equations (6a), (7) and (9)). The fractal isotherm (11) holds for any surface with $2 \leqslant D \leqslant 3$. In the next section we will derive it as special case of a more general expression.

Here we analyse the specific properties of the three isotherms (4), (4′) and (11). Because (11) does not have the deficiencies of (4) and (4′), we take it as the appropriate form of the fractal BET isotherm and consider (4) and (4′) as approximations. We denote the right-hand side of (4), (4′) and (11) by $\theta_{\mathrm{F}}(x)$, $\theta_{\mathrm{F'}}(x)$, and $\theta_{\mathrm{C}}(x)$. Numerical results for $\theta_{\mathrm{C}}(x)$ are shown in figure 1. The principal features of the three isotherms are as follows.

(a) For $D = 2$, they agree with the classical result (2) as they should. To perform the limit $D \to 2$ in (11), one integrates by parts to obtain

$$\theta_{\mathrm{C}}(x) = f_1(x) + \int_1^\infty n^{2-D}\frac{\mathrm{d}}{\mathrm{d}n}f_n(x)\,\mathrm{d}n \quad (D > 2)$$

$$\to f_\infty(x) \qquad\qquad\qquad (D \to 2). \tag{13}$$

All three isotherms are distinct for $D \neq 2$, however.

(b) $\theta_{\mathrm{F}}(x)$ factorizes into a part that depends only on c and a part that depends only on D. A similar property holds for the numerator and denominator of $\theta_{\mathrm{F'}}(x)$. It reflects the fact that the adsorption strength c is treated as originating from a global single wall. In $\theta_{\mathrm{C}}(x)$ by contrast, c enters at the level of individual pores and thus has consistently more than one wall as source. Because $\theta_{\mathrm{C}}(x)$ does not factorize, the multiple-wall effects do not average out to a single-wall description. Specifically, $\theta_{\mathrm{F}}(x)$ and $\theta_{\mathrm{F'}}(x)$ tend to overestimate the coverage. The overestimate may be as large as a factor of two because in $\theta_{\mathrm{C}}(x)$ up to one half of all walls act as 'ceiling' rather than as 'base' for multilayer growth. The overestimate

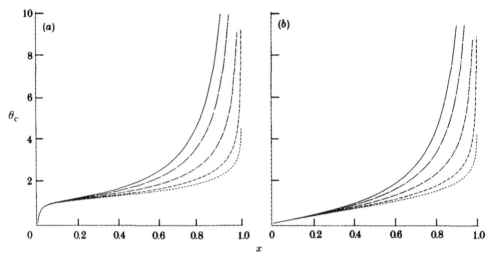

FIGURE 1. The fractal BET isotherm, equation (11). (a) The case $c = 100$ (strong adsorption). (b) The case $c = 2$ (weak adsorption). In both cases the cuves from top to bottom correspond to $D = 2.0, 2.2, 2.5, 2.8$, and 3.0.

increases with increasing surface roughness (D) and film thickness (x). Thus one expects, and indeed finds (see below), that

$$\theta_{\mathrm{F}}(x) \sim \theta_{\mathrm{F'}}(x) \sim 2\theta_{\mathrm{C}}(x) \tag{14}$$

for $D = 3$ and $x \to 1$.

(c) For $D = 3$, the isotherms $\theta_{\mathrm{F}}(x)$ and $\theta_{\mathrm{F'}}(x)$ are elementary functions:

$$\theta_{\mathrm{F}}(x) = c/(1 + (c-1)\,x)\ln\left[1/(1-x)\right], \tag{15}$$

$$\theta_{\mathrm{F'}}(x) = cx/(cx + (1-x)^2)\,(1 + \ln\left[1/(1-x)\right]), \tag{15'}$$

where use of (10) has been made. This proves the first part of (14).

(d) The low-pressure behaviour ($x \to 0$) of the isotherms is given by

$$\theta_{\mathrm{F}}(x) = c[x + (2^{2-D} - 1 - c)\,x^2 + O(x^3)] \tag{16}$$

$$\theta_{\mathrm{F'}}(x) = c[x + (2^{D-2} + 1 - c)\,x^2 + O(x^3)] \tag{16'}$$

$$\theta_{\mathrm{C}}(x) = (D-2)\int_1^\infty n^{1-D}\,c[x + (2-c)\,x^2 - (n+1)\,x^{n+1} + O(x^3)]\,\mathrm{d}n$$

$$= c[x + (2-c)\,x^2 + O(x^2/(-\ln x))]. \tag{17}$$

At very low pressures, they are linear in x and thus satisfy Henry's law. (For a general study of Henry's law on a fractal surface, see Cole *et al.* 1986.) They coincide and are independent of D in this régime, as expected. At pressures where contributions of order x^2 become important, however, the isotherms begin to differ and only $\theta_{\mathrm{C}}(x)$ is still independent of D. This is the régime of interest for determination of monolayer values N_{m} (knee in figure 1b). The fact that $\theta_{\mathrm{C}}(x)$ is unaffected by D in this régime proves the important point for applications: for arbitrary $2 \leqslant D \leqslant 3$, monolayer values N_m can be safely determined from the classical formula (2) as long as the data is restricted to the range $x < 0.15$ (figure 1).

(e) Despite the differences between (16) and (17), the isotherm $\theta_F(x)$ is a rather good approximation to $\theta_C(x)$ up to fairly high pressures. Figure 2 shows that the relative difference $(\theta_F - \theta_C)/\theta_C$ is less than 5 % for $0 \leqslant x \leqslant 0.5$ ($c = 100$) and for $0 \leqslant x \leqslant 0.8$ ($c = 2$), respectively, even in the most unfavourable case $D = 3$. Not so for the isotherm $\theta_{F'}(x)$: The difference $(\theta_{F'} - \theta_C)/\theta_C$ exceeds 20 % for $x > 0.5$ ($c = 100$) and for $x > 0.15$ ($c = 2$) in the case $D = 3$. Thus the partly corrected isotherm $\theta_{F'}(x)$ is much poorer than the uncorrected one, $\theta_F(x)$! It demonstrates the importance of multiple-wall effects. It implies that the omissions in $\theta_F(x)$ partly compensate each other and that the occurrence of $\theta_F(x) < \theta_C(x)$ for certain pressures (figure 2) presumably has no simple explanation. In view of the poor performance of $\theta_{F'}(x)$, we will no longer consider it in the sequel.

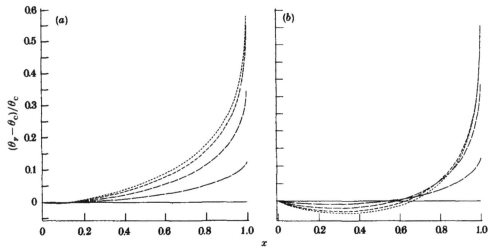

FIGURE 2. Relative difference between the isotherms (4) and (11): (a) for $c = 100$ and (b) for $c = 2$. Identically dashed curves in figure 1 and 2 have the same D value.

(f) The D dependence of $\theta_F(x)$ and $\theta_C(x)$ is described by the following results. For any fixed pressure x, the amount adsorbed per surface site decreases with increasing D in both isotherms. This is an elementary consequence of (4) and (13). It confirms that for a given number of surface sites, increasing D imposes increasing spatial restrictions on multilayer growth. For $x \to 1$, these restrictions give rise to characteristic power laws:

$$\theta_F(x) \sim \Gamma(3-D)/(1-x)^{3-D} \quad (2 \leqslant D < 3), \tag{18a}$$

$$\theta_F(x) \sim \ln(1/(1-x)) \quad (D = 3), \tag{18b}$$

$$\theta_C(x) \sim A(D)/(1-x)^{3-D} \quad (2 \leqslant D < 3), \tag{19a}$$

$$\theta_C(x) \sim (1/2)\ln(1/(1-x)) \quad (D = 3), \tag{19b}$$

where Γ is the gamma function and

$$A(D) \equiv \int_0^\infty t^{2-D} \frac{d}{dt}\left(\frac{t}{1-e^t}\right) dt \quad (2 \leqslant D < 3) \tag{20a}$$

$$= 1 \quad (D = 2). \tag{20b}$$

The prefactor $A(D)$ as a function of D is shown in figure 3. It is remarkably close to unity except for D values near 3. Equation (18a) follows from tabulated properties of the series (4) (Whittaker & Watson 1950). Equation (18b) is clear from (15). For (19), one inserts (12b) into (11) and transforms to the integration variable $t \equiv (1-x)\,n$. This gives

$$\theta_c(x) = (1-x)^{D-3}(D-2)\int_{1-x}^\infty t^{1-D}\,cx\,\frac{1-(1+t)\,e^{-u(x)\,t}}{cx(1-e^{-u(x)\,t})+1-x}\,dt \tag{21}$$

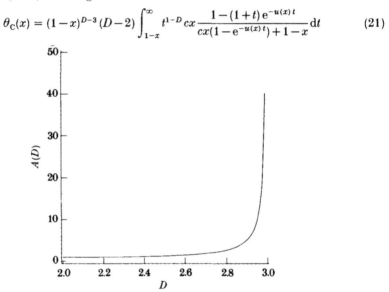

FIGURE 3. The prefactor $A(D)$ in the asymptotic relation (19a), as a function of D. The values were obtained by numerical integration of (20a)

where $u(x) \equiv (-\ln x)/(1-x)$. For $x \to 1$, one has $u(x) \to 1$ and thus

$$\theta_c(x) \sim (1-x)^{D-3}(D-2)\int_0^\infty t^{1-D}\frac{1-(1+t)\,e^{-t}}{1-e^{-t}}\,dt \quad (2 < D < 3) \tag{22a}$$

$$\theta_c(x) \sim \int_{1-x}^\infty t^{-2}\frac{1-(1+t)\,e^{-t}}{1-e^{-t}}\,dt \quad (D = 3). \tag{22b}$$

Integrating (22a) by parts gives (19a) and (20a). Writing the integrand in (22b) as $t^{-2}(\frac{1}{2}t + O(t^2\,e^{-t}))$ gives (19b). This completes the proof of equations (18)–(20). The divergence for $x \to 1$ in (18) and (19) shows that both planar and fractal surfaces adsorb infinitely thick layers as the pressure approaches saturation (on an infinitely extended fractal surface there are pores of arbitrary large size), but that the approach to infinity is increasingly slow with increasing D. For $D = 3$ the divergence is only logarithmic. For $D = 2$, (18)–(20) reproduce the classical $1/(1-x)$ divergence of (2). The fractal scaling laws (18) and (19) hold only in the régime $x \to 1$ because the adsorbed film becomes an effective prove of the fractal surface structure only when it is sufficiently thick ($\theta_C(x) \gg 1$), i.e. when the pressure is sufficiently high. We will refer to this régime as the asymptotic régime. The counterpart of thick films being sensitive to the surface geometry is that they are insensitive to the energetics in the first and second layer. This is why the adsorption strength no longer appears in (18) and (19). Thus, the BET condensation of thick films on a fractal surface has all the characteristics of a critical phenomenon: it is independent of the microscopic properties, c, of the surface; the exponents depend only on the dimensionality of the system and are insensitive to approximations (θ_F against θ_C); $D = 3$ is a critical dimension (logarithmic divergence); and the prefactors are approximation-dependent (θ_F against θ_C). The expressions (18b) and (19b) prove the factor of two in (14).

(g) There is a simple approximate argument which explains the exponent $3-D$ and the logarithm in (18) and (19) as follows. On a planar surface the mean film thickness for $x \to 1$ is ca. $a/(1-x)$ by (2). This suggests that on an arbitrary surface the film volume equals $\Omega(a/(1-x))$ where $\Omega(z)$ is the volume of all points lying at a distance not greater than z from the surface. For a D-dimensional surface and $z \gg a$, one has $\Omega(z) \sim Ba^3(z/a)^{3-D}$ where B is a unitless constant. This follows by the same argument that gave (5). For z values down to $z = a$, one has the refined equation

$$\Omega(z) = Ba^3(z/a)^{3-D} + B'a^3, \tag{23}$$

where the constant $B'a^3$ is a correction term that ensures the proper behaviour of $\Omega(z)$ at the lower end, $z = a$, of the fractal régime: $d\Omega/dz$ has the meaning of surface area as measured by molecules of diameter z and thus should equal the monolayer area $N_m a^2$ when $z = a$. At the same time, $\Omega(z)$ should equal the monolayer volume $N_m a^3$ when $z = a$. The two conditions uniquely fix the constants B and B'. The result is

$$\Omega(z) = N_m a^3 \, 1/(3-D) \, [(z/a)^{3-D} - (D-2)]. \tag{24a}$$

Clearly the correction term $(D-2)$ is negligible for sufficiently large z, as long as $D < 3$. In the limit $D \to 3$, however, the régime where this term becomes negligible is shifted to $z = \infty$ and one obtains

$$\Omega(z) \sim N_m a^3 \, 1/(3-D) \, [e^{(3-D)\ln(z/a)} - 1] \sim N_m a^3 \ln(z/a). \tag{24b}$$

Thus, substituting $z = a/(1-x)$ into (24) and converting $\Omega(z)$ into number of adsorbed molecules, one gets

$$N/N_m \sim 1/[(3-D)(1-x)^{3-D}] \quad (2 \leqslant D < 3) \tag{25a}$$

$$N/N_m \sim \ln(1/(1-x)) \qquad (D = 3) \tag{25b}$$

for $x \to 1$. This shows that the exponent $3-D$ in (18) and (19) is of purely geometric origin and that all the thermodynamics resides in the film thickness $a/(1-x)$. In fact, equations (25) are remarkably close to equations (18) (note $\Gamma(3-D) \sim$

$1/(3-D)$ for $D \to 2$ and $D \to 3$). They are off by a factor of two because the approximation $Na^3 \sim \Omega(a/(1-x))$ neglects multiple walls.

(*h*) The strong D dependence of $\theta_c(x)$ at high pressure offers a method of measuring D from one experimental isotherm, as opposed to (1), where many isotherms are needed. This may be done either by fitting the full expression (11) to the data including the low-pressure régime, or by fitting the power law (19) to high-pressure data. The first procedure has the advantage of being applicable also when data exists over a limited range only (a computer program is available upon request). The second procedure is computationally trivial. Thus, a question of some importance is how large x has to be in order for the asymptotic relations (19) to be observable. A simple estimate follows from the requirement that $\theta_c(x) \gg 1$: one requires that the right-hand side of (19) be larger than 2, which together with $A(D) \approx 1$ gives

$$x > 1 - 2^{-1/(3-D)} \tag{26}$$

as a minimal condition for x to be in the asymptotic régime. For $D = 2.5$, this yields $x > 0.75$. A maximal condition is that the right-hand side of (11) should be independent of c. This criterion can be implemented by comparing figures $1a, b$. For $D = 2.5$, it gives $x > 0.95$ if we allow the isotherms for $c = 100$ and $c = 2$ to differ by 5%. Reality lies between the minimal and maximal condition. Figure 4

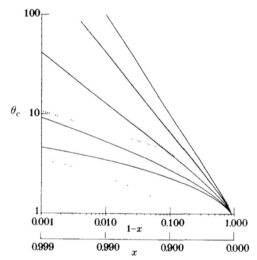

FIGURE 4 Scaling behaviour of $\theta_c(x)$ for $x > 1$. The solid lines show $\theta_c(x)$ calculated from (11) for $D = 2.0$. 2 2. 2 5. 2 8. and 3 0 (from top to bottom) The adsorption strength is $c = 100$. The dotted lines show the respective power laws (19) and have slope $D - 3$.

shows that for $D = 2.5$ the exact $\theta_c(x)$ and the power law (19) agree well when $x > 0.9$ (this amounts to $\theta_c(x) > 3$). The asymptotic régime is shifted to ever larger pressures as D goes up (figure 4) because the film volume $\Omega(z)$, (24*a*), grows ever more slowly with z. In applications one would like to convert the experimental observation of (19) over some pressure range $x_{min} \leqslant x \leqslant x_{max}$ into a statement about fractality over some length range. In the asymptotic régime the length range in question is from $a/(1-x_{min})$ to $a/(1-x_{max})$. When D is obtained from fitting (11) to data in the range $0 \leqslant x \leqslant x_{max}$, instead, the length range is about from a to $a\theta_c(x_{max})$. Useful experimental checks are to test whether D values and length ranges obtained from different adsorbates agree.

This concludes our analysis of altogether four different theories of BET adsorption on a fractal surface. We have shown that they give similar results. We have explained why we believe that (11) is to be preferred over the others. Equation (11) is not simply a power law because it includes also the crossover to the non-fractal régime ($x \to 0$). This makes (11) a particularly strong prediction. It predicts the entire fractal and non-fractal régime exclusively in terms of N_m, D and c.

Recently Ross et al. (1988) have used (11) to analyse experimental isotherms of nitrogen on fumed silica. The data extended up to $x = 0.8$ ($N/N_m = 2.7$). The best fit was obtained for $c = 150$ and $D \approx 2.5$, but deviations at high pressures were substantial. If one fits (11) to lower-pressure data only, D drops steadily to 2.0. Thus the isotherms of Ross et al. do not follow (11). The reason is quite instructive: other data by Ross et al. suggest that the surface is fractal over a relatively narrow range of length scales only. Secondly, the silica consists of compact primary particles with a diameter of at least 50 Å. The particles are therefore surface fractals and should obey the FHH isotherm, not the BET isotherm, at high pressures (table 1).

3. BET THEORY ON ARBITRARY FRACTAL SUBSTRATES

The isotherm (11) describes adsorption on a surface in three-dimensional space (topological dimension $D_{top} = 2$ of the substrate, dimension $d = 3$ of the embedding space). Here we deduce the fractal BET isotherm for arbitrary $D_{top} = 0, 1, 2, \ldots$; $d = 1, 2, 3, \ldots$; and $D_{top} \leqslant D \leqslant d$. This extension is of interest because the natural domain of the BET model at high pressure consists of mass fractals and most experimentally known mass fractals have $D_{top} = 1$ rather than $D_{top} = 2$. Indeed, most fractal aggregates, clusters and polymers have the status of curves (branched or non-branched). It will turn out that the isotherm depends both on D and D_{top}. For instance, the isotherm for $D = 2$ and $D_{top} = 1$ (random walks and lattice animals) will differ dramatically from the classical result (2). Aerogels provide examples for $D_{top} = 1$ with $D < 2$ (Schaefer et al. 1987) and $D > 2$ (Vacher et al. 1988).

Following the derivation of (11) by Cole et al. (1986), we decompose the substrate into pores of different radii R. Typically small pores are subpores of larger pores. We keep track of this by defining the space of pores with radii not greater than R as consisting of all non-substrate points that are inaccessible to balls of radius R. The d-dimensional volume of these pores with radius not greater than R scales like R^{d-D} for any fractal system, irrespective if its topological dimension (Pfeifer 1988). Thus the number of pores with radius not greater than R scales like R^{-D}. In this way we can treat the pores as independent. Moreover, in each pore we can treat the substrate as a euclidean system of dimension D_{top}

(planar surface, straight line, single point) with a $(d-1)$-dimensional ceiling at distance R. Thus the maximum number of layers that can grow in a pore of radius R equals $n = R/a$, and the number of molecules adsorbed at pressure x in the pore equals

$$M(R, x) = I(R) S(R) f_{R/a}(x). \tag{27}$$

Here $f_n(x)$ is the n-layer isotherm (12), $S(R)$ is the number of substrate sites in the pore, $I(R)$ is the number of independent columns that can grow per substrate site, and $R \geq a$. The two numbers scale like

$$S(R) \propto R^{D_{\text{top}}}, \tag{28}$$

$$I(R) \propto R^{d-1}/S(R), \tag{29}$$

(see figure 5). Substitution of (28) and (29) into (27) gives

$$M(R, x) = S(a) (R/a)^{d-1} f_{R/a}(x), \tag{30}$$

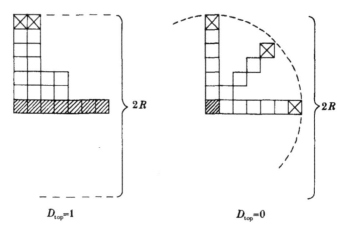

$$D_{\text{top}}=1 \qquad\qquad D_{\text{top}}=0$$

FIGURE 5 Illustration of equations (28) and (29) for $d = 2$. The filled squares represent the substrate sites (adsorption sites for the first layer). Open squares represent adsorbed molecules. Crossed squares represnt molecules at the ceiling. Their number is proportional to R^{d-1} and upon division by the number of substrate sites gives $I(R)$.

where the prefactor has been determined from the condition $I(a) = 1$, i.e. from $M(a, x) = S(a) f_1(x)$. The differential number distribution of pore radii R is given by

$$\rho(R) = \rho(a) (R/a)^{-D-1}. \tag{31}$$

Therefore the total number of adsorbed molecules at pressure x equals

$$N = \int_a^{R_{\text{max}}} M(R, x) \rho(R) \, dR = S(a) \rho(a) a \int_1^{n_{\text{max}}} n^{d-D-2} f_n(x) \, dn, \tag{32}$$

where $n_{\text{max}} = R_{\text{max}}/a$, and R_{max} is the upper limit of the fractal régime. The total number of substrate sites is

$$N_{\text{m}} = \int_a^{R_{\text{max}}} S(R) \rho(R) \, dR = S(a) \rho(a) a \int_1^{n_{\text{max}}} n^{D_{\text{top}}-D-1} \, dn \tag{33}$$

by (28). Together (32) and (33) yield

$$N/N_m = (D-D_{top})/(1-n_{max}^{-(D-D_{top})}) \int_1^{n_{max}} n^{d-D-2} f_n(x)\,dn, \quad D > D_{top}, \quad (34a)$$

$$= 1/(\ln n_{max}) \int_1^{n_{max}} n^{d-D-2} f_n(x)\,dn, \qquad\qquad D = D_{top}. \quad (34b)$$

This is the BET isotherm on an arbitrary fractal substrate in arbitrary space dimension.

We discuss various cases of (34). The result (11) for a fractal surface obtains from (34a) by setting $D_{top} = 2$, $d = 3$, and $n_{max} = \infty$. The result (2) for a planar surface obtains similarly from (34b). More generally, when $D_{top} = d-1$ (adsorption on a surface in $d = 3$, adsorption on a curve in $d = 2$, etc.) the isotherm (34) depends only on the difference $D-D_{top}$. This is in agreement with the fact that the BET equation (2) also describes adsorption on a straight line in two dimensions. For $d = 3$ and arbitrary $D_{top} = 0, 1, 2$, the thermodynamic limit of (34) yields

$$N/N_m = (D-D_{top}) \int_1^\infty n^{1-D} f_n(x)\,dn = \begin{cases} \text{finite} & \text{if } D > 2, \\ \text{finite} & \text{if } D = 2,\ D_{top} = 2, \\ \infty & \text{if } D = 2,\ D_{top} < 2, \\ \infty & \text{if } D < 2. \end{cases} \quad (35)$$

The reason for the divergence is that n^{1-D} is not integrable at infinity when $D \leqslant 2$ (the factor $f_n(x)$ is bounded by $f_1(x) \leqslant f_n(x) \leqslant f_\infty(x)$). Physically the divergence can be understood by considering the case $D = 1$: for adsorption on a straight line in $d = 3$, the film–vapour interface *increases* in area with increasing film thickness. In the BET model, this makes every finite coverage unstable with respect to further adsorption, the driving force being the gain in entropy. This gives a nominally infinite coverage at all pressures. The coverage is highly non-uniform, however, because of the infinite entropy per adsorbed molecule. It implies that the fluctuations in film thickness are infinite. This situation persists at $D > 1$ until the substrate carries sufficient ceilings to suppress infinite fluctuations. Sufficient ceilings are present when the substrate is no longer transparent, i.e. when the projection of the substrate onto a plane has positive area. By the projection theorem (see, for example, Falconer 1985) this is the case precisely when $D > 2$, in agreement with (35). Thus, $D = 2$ is a lower critical dimension below which the adsorbed phase exists only in a highly disordered state. When $D < 2$, the substrate potential is too weak to induce long-range order in the adsorbed phase.

The consequences for adsorption on fractal aggregates ($D_{top} = 1$, $d = 3$) are as follows. For $D > 2$, the isotherm is the same as for a surface, equation (11), except that the prefactor $D-2$ has to be replaced by $D-1$. Most of the previous discussion of (11) therefore applies also to aggregates. For $D \leqslant 2$, however, the isotherm exists only for finite systems, equation (34). It depends critically on the outer cut-off n_{max} of the fractal régime. Thus, while n_{max} is an optional fitting parameter when $D > 2$, it is a mandatory parameter when $D \leqslant 2$.

4. FHH THEORY ON FRACTAL SURFACES

The so-called Frenkel–Halsey–Hill theory (see, for example, Steele 1974) has been a mainstay for analysing thick-film adsorption in the case of a wetting film. Its extension to fractal surfaces (Pfeifer *et al.* 1989; Cheng *et al.* 1989), although approximate, leads to considerably less competition among different approaches than what we have seen in the BET theory. For instance, multiple-wall effects turn out to be negligible in the FHH theory. We therefore focus the discussion on the basic assumptions that enter the theory, including estimates of terms neglected in the treatment.

Even on a planar surface, there are many different derivations of the FHH theory. We present two of them from which the nature of the approximations involved will become clear. They both determine the chemical potential of the film relative to that of the bulk state of the same adsorbate. Suppose that the film, of thickness z, and the substrate are flat. If we (*a*) take the chemical potential to be the derivative of the film energy (instead of free energy) with respect to coverage, and (*b*) take the film to be structurally identical to the bulk material, then the difference $\Delta\mu$ between the chemical potentials of film and bulk is simply as a result of the replacement of a half-space of bulk by the substrate. This gives rise to a difference of potentials ΔV, which equals $\Delta\mu$ according to assumption (*a*). Since the half-spaces lie at distance z from the film surface (where an extra molecule is added), we obtain

$$\mu_{\text{film}} - \mu_{\text{bulk}} = \Delta\mu = -\alpha/z^3. \tag{36}$$

In the second relation we have used the assumption that the molecule's interaction with the surface follows an inverse distance cubed law where α is the difference of interaction constants. This is a good approximation to the van der Waals potential if z is between 5 and 100 Å. At shorter distances, overlap modifies the result; at larger z, retardation starts to reduce the potential strength.

The final (and most justified) assumption is that the vapour coexisting with the film is ideal; corrections to this can be made where appropriate. Ideality provides the difference in chemical potential of the vapour relative to saturation as

$$\Delta\mu_{\text{vapour}} = kT \ln(p/p_0) = kT \ln x. \tag{37}$$

In equilibrium, the film and vapour chemical potentials coincide, so that (36) and (37) give the same value, i.e.
$$-\alpha/z^3 = kT \ln x. \tag{38}$$

Equation (38) is the FHH relation, describing the continual growth of z as the pressure p approaches saturation p_0. Equation (36) shows that the presence of bulk solid is a crucial ingredient, and that therefore any fractal FHH theory will necessarily be restricted to surface fractals (table 1). The derivation given above is a macroscopic one; it treats the film as a continuum. Our second derivation is a microscopic one. We use a relationship derived by de Oliveira & Griffiths (1978), based on mean-field theory of a lattice–gas model:

$$m_j = \tanh\left\{(2kT)^{-1}[kT \ln x - V_j + \tfrac{1}{2}\epsilon(bm_j + b'm_{j-1} + b'm_{j+1})]\right\}. \tag{39}$$

This expresses the mean density m_j in layer j in terms of the density in neighbouring layers ($m_j = -1$: empty layer, $m_j = +1$: full layer). It takes into account the interactions among adsorbate molecules (ϵ), the interaction of the molecules with the substrate (potential $V_j = -\alpha/j^3$ for $j \geqslant 2$), and the equilibrium with the gas phase ($\ln x$). The constants b, b' are coordination numbers. We may then ask for the condition that the film–vapour interface forms at layer j. The condition is $m_{j-1} = 1$, $m_j = 0$, $m_{j+1} = -1$ and leads to $V_j = kT \ln x$, which is (38) directly. While this derivation appeals by its conceptual simplicity, it is again not free of assumptions. Here the implicit assumption is that the film density up to layer $j-1$ is the same as in bulk ($m_1 = \ldots = m_{j-1} = 1$) and then abruptly drops to gas density ($m_{j+1} = m_{j+2} = \ldots = -1$) over a distance of two layers. Thus fluctuations in film thickness are confined to a single layer. This enforces a smooth film–vapour interface (table 1).

These derivations represent the starting point for our recent work on the FHH problem for fractal substrates. The most naive calculation proceeds by considering two substrates of the same composition, one planar and the other a surface fractal. Let us calculate the coverages when they are exposed to a gas at a given pressure and temperature. The argument is that the thickness z on the planar surface, equation (38), provides the characteristic distance scale for the fractal case. That is, for the fractal substrate we construct the locus of points at distance z from the solid. This equidistance surface should be a good approximation to the equipotential surface (energy: $kT \ln x$) generated by the solid, and therefore to the film–vapour interface on the fractal. Thus we can find the coverage by calculating the volume bounded by this equidistance surface. The volume is simply $\Omega(z)$ as calculated in (23) and (24). If one neglects the term $B'a^3$ in (23) and normalizes so that $\Omega(a)$ equals the monolayer volume $N_m a^3$, one obtains

$$N/N_m = (z/a)^{3-D} = [\gamma/(-\ln x)]^{(3-D)/3}, \tag{40}$$

$$\gamma \equiv \alpha/(kTa^3), \quad 2 \leqslant D < 3, \tag{41}$$

after inserting z from (38). This is the fractal FHH isotherm as obtained in Pfeifer et al. (1989). If one uses formula (24a) for $\Omega(z)$, one gets

$$N/N_m = (3-D)^{-1}\{[\gamma/(-\ln x)]^{(3-D)/3} - (D-2)\} \tag{42}$$

instead. Both isotherms reduce to the classical FHH isotherm for $D = 2$. Both give monolayer coverage at the same D-independent pressure. The expression (42) may fit experimental data down to somewhat smaller pressures than (40) can do. In figure 6, we show the isotherm (40) for different values of D.

Clearly the derivation of (40) and (42) has conceptual input essentially identical to that of the FHH relation (38). There is the additional insight that the potential energy at a position at distance z from the fractal surface should depend only on z and is given by the flat-surface expression $-\alpha/z^3$. This assumption is equivalent to the assumption that positions with potential contributions from several nearby walls ('trough positions') are rare, i.e that multiple-wall effects are unimportant. The validity of this assumption has been checked in two distinct ways. In Pfeifer et al. (1989) we have replaced the fractal surface by an equivalent system of

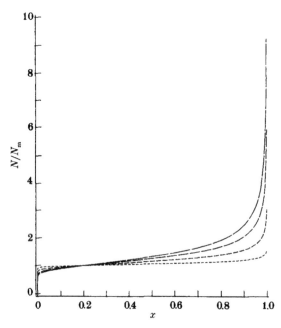

FIGURE 6. The fractal FHH isotherm, equation (40), for $D = 2.0$, 2.2, 2.5, and 2.8 (from top to bottom). The behaviour in the régime $N/N_\mathrm{m} \lesssim 1$ is physically meaningless because the potential $-\alpha/z^3$ needs to be modified for $z \lesssim a$.

independent spherical pores, similar to the treatment leading to the fractal BET isotherm (11). In each pore, we have taken the potential appropriate for a spherical cavity surrounded by an infinite solid. This explicitly incorporates multiple-wall effects; in fact, it overestimates them. The resulting isotherm is the same as (40) except for a D-dependent prefactor of 1.04 at most (for $D \approx 2.5$). This shows that the approximate potential used in deriving (40) is well justified. The same conclusion is arrived at by Cheng *et al.* (1989) who numerically computed exact equipotential surfaces for a model surface fractal and took these to be the film–vapour interface at successively higher pressures x. The computed isotherm was again in excellent agreement with the prediction (40).

The approach used in all these calculations, namely to equate the film–vapour interface to an equipotential surface (exact or approximate), can be justified from the relation (39) by de Oliveira & Griffiths: two positions with the same potential energy V_j correspond to a common interface under appropriate thermodynamic conditions. This presupposes, however, that the interface is locally flat, which need not be the case. The condition for local flatness is simply this: the radius of curvature of the interface must be large compared to the capillary length, a_c. The reader is reminded that a_c governs the general behaviour of interfaces in external fields. In the familiar case of gravity and water, a_c is a few millimetres. For adsorbed films the substrate force replaces gravity and a_c becomes of microscopic scale. An argon film has a_c comparable to the film thickness z when $z = 100$ Å; for thinner films capillary effects become unimportant. Thus for film thicknesses up to 50–100 Å, the calculations described in this section should be quite realistic.

Subsequent to the development of these fractal FHH models a set of measurements were made which conformed to the prediction. Krim found that nitrogen adsorption on a silver-plated quartz microbalance could well be fitted by (40); see figure 7. More recently, the same surface was studied by scanning

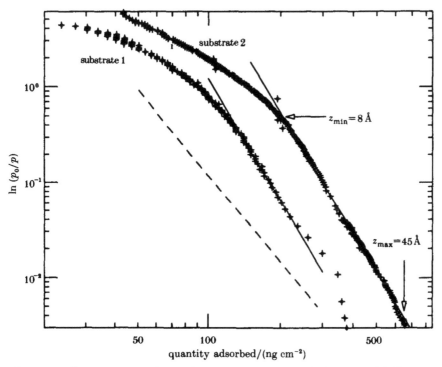

FIGURE 7. Experimental isotherms for nitrogen on silver, obeying (40) with $D = 2.30$. The highest measured pressure was $x = 0.997$, corresponding to a coverage of $N/N_m = 15$. The dashed line shows equation (40) for $D = 2.0$. (From Pfeifer *et al.* (1989).)

tunnelling microscopy, which yielded a fractal dimension consistent with the thermodynamic analysis over the same range of length scales (White *et al.* 1989). We feel that this is a rather striking example for the predictive capability of the fractal concept! From the agreement between the thermodynamic analysis and the STM analysis we can also draw a second remarkable conclusion: the silver surface, while geometrically heterogeneous ($D = 2.30$), appears to be energetically homogeneous. This conclusion obtains from the following consideration. It is known that a planar surface with energetic disorder (different constants α in different regions of the surface, cf. (36)) leads to exponents *larger* than $\frac{1}{3}$ for (40) (see, for example, Steele 1974). First, this implies that the exponent in figure 7 cannot be as a result of energetic disorder on a planar surface. Second, it suggests that for energetic disorder on a fractal surface the exponent should be larger than $\frac{1}{3}(3-D)$. But the STM result shows that the exponent equals $\frac{1}{3}(3-D)$, whence the absence of energetic disorder.

5. Conclusion

We have given an exhaustive discussion of the fractal BET theory. By doing so, we have addressed a number of questions which arise in any theory of multilayer adsorption on a fractal substrate (estimates of film volumes, presence of several nearby walls, method of pore-size distribution, onset of the asymptotic régime, crossover to D-independent behaviour at low pressures). If used as a tool to obtain monolayer values from low-pressure data, the fractal BET theory asserts that this is possible without *a priori* knowledge of the fractal nature of the substrate. If used to obtain the fractal dimension (from high-pressure data), the BET theory is limited to mass fractals, because of its restrictive physical assumptions. The physical assumptions in the FHH model, by contrast, are such that it does well where the BET model fails. As a result, the fractal FHH theory yields powerful predictions even at the stage of simple treatments as presented here. What remains to be done is fairly obvious, but non-trivial. Apart from real experiments, which are often ambiguous, an obvious test of the various assumptions is the use of computer simulations. Such simulations will allow one to examine the actual structure of the film–vapour interface, to narrow down the role of surface tension, and to explore the sensitivity to energetic disorder. The difficulty will be the much larger number of particles required than for a planar surface, and the possible importance of metastability (always a factor in porous media). A different type of extension is to go beyond physisorption, i.e. beyond van der Waals forces and the assumption of wetting. A logical generalization to such more complex systems, including mixtures, is well worth pursuing.

Acknowledgement is made to the donors of the Petroleum Research Fund, administered by the American Chemical Society, to the Weldon Spring Foundation, and to the National Science Foundation (grant DMR-8718771) for partial support of this work.

References

Avnir, D. & Pfeifer, P 1983 *Nouv J. Chim* **7**, 71–72.
Avnir, D., Farin, D. & Pfeifer, P. 1984 *Nature, Lond.* **308**, 261–263.
Avnir, D., Farin, D. & Pfeifer, P. 1984 *J Colloid Interface Sci.* **103**, 112–123.
Bale, H D. & Schmidt, P W. 1984 *Phys. Rev. Lett.* **53**, 596–599.
Berry, M. V. & Percival, I. C. 1986 *Optica Acta* **33**, 577–591.
Blender, R. & Dieterich, W. 1986 *J. Phys.* A **19**, L785–L790.
Cheng, E , Cole, M W & Stella, A 1989 *Europhys. Lett.* (In the press)
Cole, M. W., Holter, N. S. & Pfeifer, P. 1986 *Phys Rev.* B **33**, 8806–8809
de Gennes, P. G. 1982 *C. r. Acad Sci., Paris* I **295**, 1061–1064
de Oliveira, M. J. & Griffiths, R. B 1978 *Surf. Sci.* **71**, 687–694
Even, V., Rademann, K , Jortner. J . Manor. N. & Reisfeld, R. 1984 *Phys. Rev. Lett.* **52**, 2164–2167
Falconer, K J. 1985 *The geometry of fractal sets.* Cambridge University Press.
Farin, D & Avnir, D. 1989 In *The fractal approach to heterogeneous chemistry* (ed. D. Avnir). Chichester Wiley & Sons (In the press.)

Frank, H., Zwanziger, H. & Welsch, T. 1987 *Z. analyt Chem.* **326**, 153–154.
Fripiat, J. J. 1989 In *The fractal approach to heterogeneous chemistry* (ed. D. Avnir). Chichester: Wiley & Sons. (In the press.)
Fripiat, J. J., Gatineau, L. & Van Damme, H. 1986 *Langmuir* **2**, 562–567.
Klafter, J. & Blumen, A. 1984 *J. chem. Phys.* **80**, 875–877.
Levitz, P., Van Damme, H. & Fripiat, J. J. 1988 *Langmuir* **4**, 781–782.
Liu, S. H. 1985 *Phys. Rev. Lett.* **55**, 529–532.
Martin, J. E. & Hurd, A. J. 1987 *J. appl. Crystallogr.* **20**, 61–78.
Nyikos, L. & Pajkossi, T. 1985 *Electrochim. Acta* **30**, 1533–1540.
Nyikos, L. & Pajkossi, T. 1986 *Electrochim. Acta* **31**, 1347–1350.
Pfeifer, P. 1987 In *Preparative chemistry using supported reagents* (ed. P. Laszlo), pp. 13–33. San Diego: Academic Press.
Pfeifer, P. 1988 In *Chemistry and physics of solid surfaces* VII (ed. R. Vanselow & R. F. Howe), pp. 283–305. Berlin: Springer.
Pfeifer, P. & Avnir, D. 1983 *J. chem. Phys.* **79**, 3558–3565.
Pfeifer, P., Avnir, D. & Farin, D. 1984 *J. statist. Phys.* **36**, 699–716.
Pfeifer, P., Wu, Y. J., Cole, M. W. & Krim, J. 1989 *Phys. Rev. Lett.* (Submitted.)
Pines, D., Huppert, D. & Avnir, D. 1988 *J. chem. Phys.* **89**, 1177–1180.
Ross, S. B., Smith, D. M., Hurd, A. J. & Schaefer, D. W. 1988 *Langmuir* **4**, 977–982.
Schaefer, D. W., Wilcoxon, J. P., Keefer, K. D., Bunker, B. C., Pearson, R. K., Thomas, I. M. & Miller, D. E. 1987 In *Physics and chemistry of porous media* II (ed. J. R. Banavar, J. Koplik & K. W. Winkler), pp. 63–80. New York: American Physical Society.
Schmidt, P. W. 1988 *Makromol. Chem., Macromol. Symp.* **15**, 153–166.
Spindler, H., Szaragan, P. & Kraft, M. 1987 *Z. Chem.* **27**, 230–231.
Steele, W. A. 1974 *The interaction of gases with solid surfaces*. Oxford: Pergamon Press.
Vacher, R., Woignier, T., Pelous, J. & Courtens, E. 1988 *Phys. Rev.* B **37**, 6500–6503.
Vlachopoulos, N., Liska, P., Augustynski, J. & Grätzel, M. 1988 *J. Am. chem. Soc.* **110**, 1216–1220.
White, H. W., Wragg, J. L., West, J., Kenntner, J. & Pfeifer, P. 1989 Unpublished.
Whittaker, E. T. & Watson, G. N. 1950 *A course in modern analysis*, p. 280. Cambridge University Press.
Wong, P. Z., Howard, J. & Lin, J. S. 1986 *Phys. Rev. Lett.* **57**, 637–640.

Reactions in and on fractal media

By A. Blumen and G. H. Köhler

Physical Institute and BIMF, University of Bayreuth, D-8580 Bayreuth, F.R.G.

This article deals with bimolecular chemical reactions. Whereas the dynamics under well-stirred conditions is readily described by ordinary differential equations, to account for geometrical or energetic restrictions is a much harder task. In the framework of random walks we discuss some modern approaches of treating different disorder aspects. We focus on fractals, which provide a good picture for spatial randomness, and on ultrametric spaces, which mimic energetic disorder. Furthermore, differences in waiting times can be incorporated in the general formalism in terms of continuous-time random walks. The study of the survival probability of the chemical species leads to rich temporal behaviours, as for instance to stretched exponential or to algebraic decay patterns. We point out the importance of the mean number of distinct lattice points visited in time t, and give a derivation of this quantity for regularly multifurcating ultrametric spaces.

1. Introduction

Randomness occurs in many areas of modern physics. For example, in the past decade interest has turned increasingly towards the investigation of amorphous solids such as glasses. There are two reasons why this did not happen earlier: on the one hand these substances are now of technical interest; on the other hand, until recently the lack of suitable theoretical methods prevented a satisfactory description of experimental results. It is clear that models developed for crystalline solids are inappropriate, because they are based upon the translational symmetry of the lattice, and glasses are translationally disordered. Glasses show a number of unusual features, such as anomalies in diffusion (Schnörer *et al.* 1988) and specific heat (Pohl 1981; Stephens 1976) as well as ultra-sound absorption (Jäckle 1972) and photochemical hole-burning (W. Köhler & Friedrich 1987).

Another area in which randomness is central is that of reactions under diffusion-limited conditions, a classical topic in physical chemistry (Chandrasekhar 1943; Calef & Deutch 1983). New analytical methods and increased computing power have demonstrated new, unexpected features (Balagurov & Vaks 1973; Donsker & Varadhan 1979; Toussaint & Wilczek 1983; Zumofen *et al.* 1985). These investigations have revealed many qualitative deviations from the accepted Smoluchowski-type decay laws, when reactions in confined geometries, such as obtained for limited dimensionalities or for porous media, are studied.

Randomness can occur in many forms. Thus, transport properties of spatially random systems (mixed crystals, alloys) are triggered by a distribution of microscopic (site-to-site) transfer rates (temporal disorder) and by different interactions with the surroundings (energetic disorder). In this article we present several classes of models for disorder, which have come under close scrutiny only during the past decade. Each class may be viewed as arising from a particular aspect of randomnes. Fractals may be looked upon as a model for spatial disorder (Mandelbrot 1982; Falconer 1985), whereas continuous-time processes (Scher & Lax 1973 a, b; Scher & Montroll 1975; Pfister & Scher 1978; Montroll & Shlesinger 1984) exemplify the temporal, and ultrametric structures (Bourbaki 1974; Schikhof 1984; Rammal et al. 1986) the energetic disorder.

This paper focuses on the kinetics of the $A + A \to 0$ and the $A + B \to 0$ diffusion-limited reactions by modelling the dynamics through random walks. The disorder aspect is depicted through the hierarchical (self-similar) structures already discussed; as a consequence we will often find temporal decay patterns that differ from the usual kinetic scheme.

The structure of the paper is as follows. In §2 we give a summary of basic features of random walks (RW) and continuous-time random walks (CTRW), and present the connection between them using generating functions. In §3 we recall the results of chemical kinetics and display deviations that occur in random media. In §4 diffusion is taken into account by using the RW formalism both on fractals and on ultrametric spaces (UMS); we use the target model to compute the decay of the minority species. The result will show the importance of S_n, the mean number of distinct sites visited in n steps, and of $S(t)$, the analogous function for continuous time. This is also stressed by the findings of §5, where the $A + A \to 0$ reaction is also governed by $S(t)$. Because of the fundamental role of $S(t)$, we devote §6 to its derivation on UMS under CTRW-conditions. Comparison to former results shows the close relation between UMS and fractals. Section 7 contains our conclusions.

2. Connection between random walks and continuous-time random walks; generating functions

In this article we will describe the dynamics in terms of random walks (RW), so that the motion takes place on discrete lattices. Important types of lattices are regular ones (like the simple cubic lattice), but also fractals, ultrametric spaces (UMS) and related concepts (the Knapp-space (KS), and Cayley trees). In general, one distinguishes between two possibilities.

Firstly, the simple RW, which perform steps with a fixed frequency. A new step occurs exactly after a characteristic time τ has elapsed. Expressions of interest are for example $P_n(0)$, the probability that the RW has returned to the origin 0 in the nth step; S_n, the mean number of distinct sites visited in n steps; or Φ_n, the probability that a particle has survived in n steps.

Secondly, the continuous-time random walks (CTRW). Now the condition that steps may occur only at preassigned times is relaxed. Instead of this, one introduces a waiting-time distribution $\psi(t)$, which gives the probability density that the time between steps equals t. A common example is the so-called Poisson process where $\psi(t)$ is of exponential form, $\psi(t) = \lambda \exp(-\lambda t)$. The λ can be interpreted as the (constant) rate of leaving a site. In general, so-called Markov processes (i.e. processes that can be described by a master equation) are always of this form, but with rates $\lambda(x)$, which may depend on the site x.

Another important example for $\psi(t)$ is from Scher & Montroll (1975) who used forms that display long-time tails:

$$\psi(t) \sim t^{-1-\alpha}. \tag{1}$$

Interestingly, equation (1) is intimately related to a fractal set of event times (Montroll & Shlesinger 1984; Shlesinger 1984; Blumen *et al.* 1986*a*, *b*). Indeed, by taking into account events on all timescales, one readily constructs

$$\psi(t) = \frac{1-N}{N} \sum_{j=1}^{\infty} N^j q^j \exp\left(-tq^j\right), \tag{2}$$

where $N < 1$. As is evident, the distribution (2) is (approximately) dilatationally invariant, because

$$\psi(qt) = \frac{\psi(t)}{Nq} - \left(\frac{1-N}{N}\right) \exp\left(-tq\right)$$

and thus at longer times: $\psi(qt) = \psi(t)/(Nq)$ for $q < N$, i.e. $q < 1$. Hence equation (2) is equivalent to equation (1) when one sets $\alpha = \ln N / \ln q$. Equation (1) shows directly the temporal scaling of $\psi(t)$, i.e. its fractal nature in time.

The connection between RW and CTRW was implemented in a classic work by Montroll & Weiss (1965), by using generating functions. We exemplify the procedure with $P_n(0)$, because the derivations for S_n and Φ_n are quite analogous.

Thus let $\tilde{P}_0(z)$ be the generating function for being at the origin 0. $\tilde{P}_0(z)$ is defined through the series

$$\tilde{P}_0(z) \equiv \sum_{n=0}^{\infty} P_n(0) z^n. \tag{3}$$

To obtain $\tilde{P}_0(z)$, we follow Montroll & Weiss (1965), Scher & Montroll (1975), Blumen & Zumofen (1982), and Köhler & Blumen (1987). Let $\phi_n(t)$ be the probability of having performed exactly n steps in time t. Then

$$P_0(t) = \sum_{n=0}^{\infty} \phi_n(t) P_n(0). \tag{4}$$

Here $P_0(t)$ denotes the probability to be at 0 after the time t has elapsed, which is the continuous time analogue of $P_n(0)$ for discrete time. We now have to express $\phi_n(t)$ through $\psi(t)$, the waiting-time distribution between steps. Writing $\hat{f}(u) \equiv \mathscr{L}[f(t)]$ where $\mathscr{L}$ is the Laplace transform, one has (Blumen & Zumofen 1982):

$$\hat{\phi}_n(u) = \frac{1-\hat{\psi}(u)}{u} [\hat{\psi}(u)]^n.$$

Thus Laplace transformation of equation (4) yields

$$\hat{P}_0(u) = \frac{1-\hat{\psi}(u)}{u} \sum_{n=0}^{\infty} [\hat{\psi}(u)]^n P_n(0), \tag{5}$$

which, apart from the factor $[1-\hat{\psi}(u)]/u$, is nothing more than the generating function (3) of the $P_n(0)$ evaluated at $z = \hat{\psi}(u)$ (Montroll & Weiss 1965). Thus the connection between RW and CTRW is most transparent in the Laplace domain, when use is made of generating functions. As an important result, from equation (5) one may switch from $P_n(0)$ to $\tilde{P}_0(z)$ to $P_0(t)$ if only one of them is known.

To end this section we note a remarkable connection between $\tilde{P}_0(z)$ and $\tilde{S}(z)$ that was found by Montroll & Weiss (1965) for regular lattices:

$$(1-z)\tilde{S}(z) = [(1-z)\tilde{P}_0(z)]^{-1}. \tag{6}$$

We were able (Köhler & Blumen 1987) to show that equation (6) is valid for arbitrary homogeneous lattices, i.e. for lattices where all points are equivalent. We will use equation (6) in §6 to determine $S(t)$ both on regularly multifurcating UMS and on the KS. Finally, in the Laplace domain it follows from equations (5) and 6) that

$$u\hat{S}(u) = 1/[u\hat{P}_0(u)]. \tag{7}$$

From this discussion it should by now be obvious that we may obtain both $P_0(t)$ and $S(t)$ for general $\psi(t)$ if either $P_0(t)$ or $S(t)$ is known for a specific $\psi(t)$, or, put differently, one special solution suffices.

3. The basic kinetic approach

In this section we recall some basic notions of chemical kinetics (Blumen *et al.* 1986*a*). An irreversible bimolecular reaction $A + B \xrightarrow{k} 0$ is described under well-stirred conditions by the following set of nonlinear differential equations:

$$dA(t)/dt = -kA(t)B(t) = dB(t)/dt. \tag{8}$$

Here $A(t)$ and $B(t)$ are the concentrations of species A and B. With the initial conditions $A_0 \equiv A(0), B_0 \equiv B(0)$, the solution of equation (8) is given by

$$[1+C/A(t)]/(1+C/A_0) = e^{Ckt}, \tag{9}$$

where $C \equiv B_0 - A_0$. From equation (9) we infer for $B_0 \gg A_0$, that $C \approx B_0$ and thus $C/A(t) \gg 1$:

$$A(t) \approx A_0 e^{-B_0 kt}. \tag{10}$$

Thus the decay of the minority species is quasi-exponential.

On the other hand, if $A_0 = B_0$ then $C = 0$ in equation (9). An expansion in small C leads to the decay

$$A(t) = A_0/(1+A_0 kt) \tag{11}$$

from which at long times an algebraic time dependence emerges:

$$A(t) \sim 1/kt. \tag{12}$$

In the special case $A = B$, i.e.

$$\tfrac{1}{2}[dA(t)/dt] = -k[A(t)]^2, \tag{13}$$

separation of the variables and integration lead to

$$A(t) = A_0/(1+2A_0 kt) \tag{14}$$

a form akin to equation (11). The long-time behaviour obeys here

$$A(t) \sim 1/2kt. \tag{15}$$

Thus, from unimolecular and from bimolecular reactions one has as long-time decays either exponential or $1/t$ algebraic dependencies.

In random materials, however, more complex decay patterns may occur. As examples we mention the Kohlrausch–Williams–Watts (stretched exponential) law

$$\Phi(t) = \exp\left[-(t/\tau)^{\beta}\right] \tag{16}$$

(Kohlrausch 1847; Williams & Watts 1970; for experimental results see Plonka *et al.* 1979; Richert & Bässler 1985), and algebraic decays

$$\Phi(t) = (t/\tau)^{-\xi} \tag{17}$$

as reported by Weiss (1982–83) and Tauc (1984). Thus simple chemical kinetics is not always sufficient to describe the experimental data. In the next sections we extend therefore the reaction–diffusion mechanisms by using random walks on different model structures.

4. The target model

In the next two sections we discuss two RW reaction schemes. We begin with the framework of the pseudo-unimolecular reactions, in which we have a minority and a majority species. In the simplest models one has one A and several B particles, and the A particle is annihilated on encounter with a B particle. Depending on which of the species performs the motion one distinguishes between several models. Here we restrict ourselves to the target model, where only the B moves (Zumofen *et al.* 1985; Blumen *et al.* 1986a). We monitor the survival probability of the A particle averaged over all possible realizations of particle distributions and motions. Interestingly, this model can be solved exactly, both for regular lattices (Blumen *et al.* 1984b) and also for ultrametric spaces (Köhler & Blumen 1987). We focus on the fate of an immobile A molecule, which gets annihilated by the mobile B species. At the start, the non-interacting B molecules are distributed randomly, with probability p over the structure.

For the survival probability Φ_n of the A molecule in a n-step walk one obtains

$$\Phi_n = \exp\left[-p(S_n - 1)\right] \sim \exp\left(-pS_n\right). \tag{18}$$

Now, for a one-dimensional lattice one has $S_n \sim n^{\frac{1}{2}}$ for large n, whereas for three dimensions the increase of S_n is linear in n: $S_n \sim n$. Hence, with equation (18) this model yields for $d = 1$ a stretched exponential decay, equation (16), with $\beta = \frac{1}{2}$.

Moreover, equation (18) holds approximately for fractals. In this case one has for S_n:

$$S_n \sim \begin{cases} n^{\frac{1}{2}\tilde{d}} & \text{for } \tilde{d} < 2, \\ n & \text{for } \tilde{d} > 2, \end{cases} \tag{19}$$

where $\tilde{d}$, not necessarily integer, is the spectral dimension of the fractal. The results for regular lattices carry over to the more general form

$$\Phi_n \approx \begin{cases} \exp\left(-Cpn^{\frac{1}{2}\tilde{d}}\right) & \text{for } \tilde{d} < 2, \\ \exp\left(-Cpn\right) & \text{for } \tilde{d} > 2. \end{cases} \tag{20}$$

We note that for $\tilde{d} < 2$, equation (20), provides a way of experimentally determining the spectral dimension of fractal objects, by monitoring the relaxation in such materials under RW conditions (Blumen *et al.* 1984b).

Finally, similar results are also valid for other hierarchical models such as UMS. Here we present the essential results; for a detailed derivation see Köhler & Blumen (1987).

We start by recalling that a UMS with branching ratio b consists of clusters of b points. Each pair of points belonging to the same cluster is assigned a distance of one; b of those clusters are gathered up in a cluster of clusters, and points belonging to different subclusters are assigned a distance of two. Continuation of this procedure leads to a hierarchical structure of clusters consisting of b subclusters each. The UMS consists of all points belonging to the lowest hierarchy, and between two points there exists by construction a uniquely determined integer distance. The points together with the distance form a so-called ultrametric space (UMS). The distance m between two points corresponds to the energy barrier separating them. We thus have here restricted ourselves to the case of linearly increasing barriers, where the spacing between successive hierarchies is a constant, Δ. The walk on the UMS is taken to be thermally activated, so that the rate ϵ_{ij} of getting from i to j is proportional to the Boltzmann factor $R = \exp(-\Delta/kT)$:

$$\epsilon_{ij} = \tau^{-1} b^{-m} e^{-m\Delta/kT} \equiv (R/b)^m / \tau. \tag{21}$$

Here τ is a characteristic time. The analysis shows that for Markov processes, i.e. for $\psi(t) = \lambda \exp(-\lambda t)$, at longer times the following relations hold:

$$P_0(t) \sim t^{-\gamma} \tag{22}$$

and

$$S(t) \sim \begin{cases} t^\gamma & \text{for } \gamma < 1, \\ t & \text{for } \gamma > 1, \end{cases} \tag{23}$$

where

$$\gamma \equiv (\ln b) kT/\Delta. \tag{24}$$

From equations (19) and (23) one infers that the spectral dimension $\tilde{d}$ of the fractal may be related to the UMS parameter γ through $\tilde{d} = 2\gamma$.

Let us now turn to the CTRW case. A detailed study (Zumofen & Blumen 1982) shows that the RW result carries over to the CTRW result:

$$\Phi(t) \sim \exp[-pS(t)]. \tag{25}$$

We note that here $S(t)$ depends on the choice of the waiting-time distribution $\psi(t)$.

For exponential waiting-time distributions, $\psi(t) = \lambda \exp(-\lambda t)$, the S_n and $S(t)$ show the same asymptotic behaviour for large arguments. Hence in equations (19) and (20) n can be simply replaced by t. The situation changes when one uses the algebraic waiting-time distributions of equation (1), $\psi(t) \sim t^{-1-\alpha}, 0 < \alpha < 1$. Here for regular lattices one finds

$$S(t) \sim t^\alpha$$

and, as an extension, for fractals:

$$S(t) \sim \begin{cases} t^{\frac{1}{2}\alpha \tilde{d}} & \text{for } \tilde{d} < 2, \\ t^\alpha & \text{for } \tilde{d} > 2. \end{cases} \tag{26}$$

In equation (26) the two fractal exponents (α for the temporal and $\frac{1}{2}\tilde{d}$ for the spatial aspect) combine multiplicatively, i.e. the two processes subordinate

(Klafter *et al.* 1984; Blumen *et al.* 1984*a*). Furthermore, for UMS and algebraic waiting-time distributions $\psi(t) \sim t^{-1-\alpha}$ one has (Köhler & Blumen 1987):

$$S(t) \sim \begin{cases} t^{\alpha\gamma} & \text{for} \quad \gamma < 1, \\ t^{\alpha} & \text{for} \quad \gamma > 1, \end{cases} \tag{27}$$

and
$$S(t) \sim 1/P_0(t). \tag{28}$$

Equation (26) parallels the UMS-case, equation (27), when one sets $\tilde{d} \equiv 2\gamma$.

More detailed investigations support the validity of equation (25) for algebraic waiting-time distributions $\psi(t)$. Because α always lies in the range 0 to 1, one obtains a stretched exponential decay, equation (16), for $\Phi(t)$.

5. THE BIMOLECULAR REACTION $A + A \to 0$

Next we centre on the bimolecular reaction $A + A \to 0$ and reference the study of the more complex $A + B \to 0 \, (A_0 = B_0)$ reaction. An overview of the whole topic is given by Blumen *et al.* (1988). As discussed in §3, the chemical kinetic scheme for the $A + A \to 0$ and for the strictly bimolecular $A + B \to 0 \, (A_0 = B_0)$ reactions, in both cases under well-stirred conditions, decay with a $1/t$ dependence. From the previous study of pseudo-unimolecular reactions at longer times, we found that the kinetic exponential was modified by the appearance of $S(t)$. Thus one might expect that the $A + A \to 0$ decay will follow a $1/S(t)$ law at longer times. For RW this corresponds to

$$\Phi_n^{AA} \sim \begin{cases} n^{-\frac{1}{2}\tilde{d}} & \text{for} \quad \tilde{d} < 2, \\ n^{-1} & \text{for} \quad \tilde{d} > 2. \end{cases} \tag{29}$$

Numerical simulations established that equation (29) describes the decay behaviour correctly both for regular lattices and for fractals (Blumen *et al.* 1988). In the CTRW scheme $S(t)$ replaces S_n so that one should find

$$\Phi^{AA}(t) \sim \begin{cases} t^{-\frac{1}{2}\alpha\tilde{d}} & \text{for} \quad \tilde{d} < 2, \\ t^{-\alpha} & \text{for} \quad \tilde{d} > 2. \end{cases} \tag{30}$$

Numerical simulations support equation (30) well (Blumen *et al.* 1986*b*), so that all findings are consistent with $\Phi^{AA}(t) \sim [S(t)]^{-1}$. Equation (30) is then another example of subordination.

To conclude our study of the $A + A \to 0$ reaction we also consider UMS. Paralleling the previous discussion, we expect at longer times

$$\Phi^{AA}(t) \sim \begin{cases} t^{-\gamma} & \text{for} \quad \gamma < 1, \\ t^{-1} & \text{for} \quad \gamma > 1. \end{cases} \tag{31}$$

This is in good agreement with numerical studies (Blumen *et al.* 1988). Thus, on UMS the trend found for regular lattices and for fractals continues: Φ_n^{AA} may be well approximated through $\Phi_n^{AA} \sim S_n^{-1}$. Hence, in all cases investigated, the long-time decay Φ_n^{AA} follows an algebraic form $\Phi_n^{AA} \sim n^{-\xi}$ (with $\xi = \gamma$ for $2\gamma = \tilde{d} < 2$, and $\xi = 1$ for $2\gamma = \tilde{d} > 2$).

6. CTRW ON REGULAR MULTIFURCATING UMS

The findings of the last sections have stressed the importance of S_n and $S(t)$, the mean number of distinct sites visited by a RW. Our results for CTRW on UMS, equations (23) and (27), turn out to be very similar to equations (19) and (26) for fractals. We devote this section to the evaluation of S_n and $S(t)$ for UMS, and follow the procedure in Köhler & Blumen (1987). The derivation is remarkable, in that it is very general. Thus it carries over without major change to other homogeneous lattices such as the Knapp-space KS (Knapp 1988; Köhler & Blumen 1988).

The starting point of the investigation is the knowledge of a particular solution for $P_0(t)$. In the Markov case the hopping rates ϵ_m are given by equation (21), see §4, and the rate λ of leaving a site i is

$$\lambda \equiv \sum_{j \neq i} \epsilon_{ij}. \tag{32}$$

Because of the homogeneity of the model λ is independent of the particular choice of i. Thus the Markov process with constant λ is a Poisson process with waiting-time distribution $\psi(t) = \lambda \exp(-\lambda t)$.

Bachas & Huberman (1986) solved the master equation for this problem and found for $P_0(t)$ that

$$P_0(t) = (b-1) \sum_{m=1}^{\infty} b^{-m} \exp\left(-R^m \frac{t}{\tau} \frac{b-R}{b(1-R)}\right) \tag{33}$$

so that one particular solution is known. From equation (33) it is a simple matter to show that the scaling relation $P_0(Rt) \sim bP_0(t)$ holds, from which $P_0(t) \sim t^{-\gamma}$, equation (22) follows, where γ is given by equation (24). Using the general CTRW theory of §2, from equations (3), (5) and (33), one concludes that the generating function of the walk is given by (see Köhler & Blumen 1987)

$$\tilde{P}_0(z) = (b-1) \sum_{m=1}^{\infty} b^{-m}(1-z+CzR^m)^{-1}, \tag{34}$$

with $C = (b-R)/(bR-R)$. For this to be true it is only necessary to observe that the probability of leaving a site is given by $\exp(-\lambda t)$, with λ being

$$\lambda = \frac{1}{\tau} \sum_{m=1}^{\infty} (b^m - b^{m-1})(R/b)^m = R(b-1)/(b-bR)\tau.$$

The corresponding waiting time distribution is $\psi(t) = \lambda \exp(-\lambda t)$ or, in the Laplace domain, $\hat{\psi}(u) = \lambda/(u+\lambda)$. Laplace transforming equation (33) and setting $z \equiv \hat{\psi}(u) = \lambda/(u+\lambda)$, i.e. $u = \lambda(1-z)/z$, leads to (34). Equation (34), together with (6), lead immediately to $\tilde{S}(z)$.

We can now easily retrieve $S(t)$ in the Markov case, by using $P_0(t) \sim t^{-\gamma}$ together with equation (7). This implies

$$\hat{S}(u) \sim \begin{cases} u^{-\gamma-1} & \text{for } \gamma < 1, \\ u^{-2} & \text{for } \gamma > 1, \end{cases}$$

from which, using Tauberian theorems (Feller 1971), one has in the time domain

$$S(t) \sim \begin{cases} t^{\gamma} & \text{for} \quad \gamma < 1, \\ t & \text{for} \quad \gamma > 1. \end{cases} \tag{35}$$

This is equation (23) of §4.

Let us now turn to CTRW on UMS. The knowledge of the generating function $\tilde{P}_0(z)$, equation (34), allows us to handle the case of general CTRW on UMS. We now set $\psi(t) \sim t^{-1-\alpha} (0 < \alpha < 1)$ so that the small-u behaviour is given by $\hat{\psi}(u) \sim 1 - cu^{\alpha}$. Insertion of $\hat{\psi}(u) = z$ in (34) leads with equation (5) to

$$\hat{P}_0(u) = \frac{1 - \hat{\psi}(u)}{u} (b-1) \sum_{m-1}^{\infty} b^{-m} \frac{1}{1 - \hat{\psi}(u) + \hat{\psi}(u) \, CR^m}$$

$$\sim u^{\alpha-1} \sum_{m-1}^{\infty} b^{-m} \frac{1}{u^{\alpha} + CR^m},$$

from which follows (Köhler & Blumen 1987)

$$\hat{P}_0(u) \sim \begin{cases} u^{\alpha-1} u^{\alpha(\gamma-1)} = u^{\alpha\gamma-1} & \text{for} \quad \gamma < 1, \\ u^{\alpha-1} & \text{for} \quad \gamma > 1. \end{cases}$$

In the long-time limit this corresponds to

$$P_0(t) \sim \begin{cases} t^{-\alpha\gamma} & \text{for} \quad \gamma < 1, \\ t^{-\alpha} & \text{for} \quad \gamma > 1. \end{cases} \tag{36}$$

Finally, the behaviour of $S(t)$ under CTRW conditions follows readily by applying relation (7) to (36). For $1 - \hat{\psi}(u) \sim u^{\alpha}$ one obtains in the Laplace domain

$$\hat{S}(u) \sim \begin{cases} u^{-\alpha\gamma-1} & \text{for} \quad \gamma < 1, \\ u^{-\alpha-1} & \text{for} \quad \gamma > 1, \end{cases}$$

and hence

$$S(t) \sim \begin{cases} t^{\alpha\gamma} & \text{for} \quad \gamma < 1, \\ t^{\alpha} & \text{for} \quad \gamma > 1, \end{cases} \tag{37}$$

which is equation (27) of §4. We notice that for $\alpha < 1$ and $\gamma < 1$ the two parameters combine multiplicatively in the exponent, hence the two processes subordinate. This is completely analogous to the findings for fractals, equation (26), if one puts $\tilde{d} = 2\gamma$. In contrast to the fractal case, however, here one of the parameters, namely γ, depends on the temperature, see (24). This fact provides a way to decide experimentally whether the temporal (α) or the energetic (γ) aspect is dominant, by performing measurements at different temperatures.

7. CONCLUSIONS

Firstly, the target model on hierarchical structures leads to stretched exponential decays. This is because of the fact that $\Phi(t) \sim \exp[-CpS(t)]$ and that $S(t)$, the mean number of distinct sites visited by the walk, may increase sublinearly with t. For fractals this happens when $\tilde{d}$, the spectral dimension, is smaller than two. On the other hand for values $\tilde{d}$ greater than two $S(t)$ increases linearly with t. For CTRW with algebraic waiting-time distributions and $\alpha < 1$, the behaviour of $S(t)$ stays in the sublinear range for all $\tilde{d}$. However, for $\tilde{d} < 2$ the exponent of $S(t)$ is given by $\frac{1}{2}\alpha\tilde{d}$ (subordination), whereas for $\tilde{d} > 2$ the temporal behaviour is governed by α alone. Furthermore for UMS one obtains results analogous to those for fractals, where however $\tilde{d}$ is replaced by 2γ.

Secondly, the findings carry over to the $A + A \rightarrow 0$ reaction. Here, the survival probability obeys the law $\Phi(t) \sim 1/S(t)$, and the decay behaviour is again dictated by $S(t)$.

Thus, in all cases, fractals and UMS behave similarly, when one identifies $\tilde{d}$ with 2γ. Because γ depends on the temperature, whereas $\tilde{d}$ and α are not sensitive to temperature changes, one has a convenient way of differentiating experimentally between the energetic and the other aspects of disorder.

We thank Professor P. Argyrakis, Professor J. Friedrich, Professor D. Haarer, Professor J. Klafter, Professor E. W. Knapp and Professor G. Zumofen for helpful discussions. The support of the Deutsche Forschungsgemeinschaft (SFB 213) and of the Fonds der Chemischen Industrie are gratefully acknowledged.

REFERENCES

Bachas, C. P. & Huberman, B. A. 1986 *Phys. Rev. Lett.* **57**, 1965.
Balagurov, B. Ya. & Vaks, V. G. 1973 *Zh. éksp. teor. Fiz.* **65**, 1939 (English translation 1974 *Soviet Phys. JETP* **38**, 968.)
Blumen, A., Klafter, J., White, S. & Zumofen, G. 1984a *Phys. Rev. Lett.* **53**, 1301.
Blumen, A., Klafter, J. & Zumofen, G. 1986a In *Optical spectroscopy of glasses* (ed. I. Zschokke), pp. 199–265. Dordrecht: D. Reidel.
Blumen, A., Klafter, J. & Zumofen, G. 1986b In *Fractals in physics* (ed. L. Pietronero & E. Tossatti), p. 399. Amsterdam: North Holland.
Blumen, A. & Zumofen, G. 1982 *J. chem. Phys.* **77**, 5127.
Blumen, A., Zumofen, G. & Klafter, J. 1984b *Phys. Rev.* B**30**, 5379.
Blumen, A., Zumofen, G. & Klafter, J. 1988 In *Fractals, quasicrystals, chaos, knots and algebraic quantum mechanics* (ed. A. Amann *et al.*), pp. 21–52. Dordrecht: Kluwer Academic Publishers.
Bourbaki, N. 1974 *Eléments de mathématique, topologie générale*, ch. 9. Paris: CCLS.
Calef, D. F. & Deutch, J. M. 1983 *A. Rev. phys. Chem.* **34**, 493.
Chandrasekhar, S. 1943 *Rev. mod. Phys.* **15**, 1.
Donsker, M. D. & Varadhan, S. R. S. 1979 *Communs pure appl. Math.* **32**, 721.
Falconer, K. J. 1985 *The geometry of fractal sets.* Cambridge University Press.
Feller, W. 1971 *An introduction to probability theory and its applications*, vol. 2. New York: Wiley.
Jäckle, J. 1972 *Z. Phys.* **257**, 212.
Klafter, J., Blumen, A. & Zumofen, G. 1984 *J. statist. Phys.* **36**, 561.
Knapp, E. W. 1988 *Phys. Rev. Lett.* **60**, 2386.
Kohlrausch, R. 1847 *Annln Phys.* **12**, 393.
Köhler, G. & Blumen, A. 1987 *J. Phys.* A**20**, 5627.
Köhler, G. H., Knapp, E. W. & Blumen, A. 1988 *Phys. Rev.* B **38**, 6774.
Köhler, W. & Friedrich, J. 1987 *Phys. Rev. Lett.* **59**, 2199.
Mandelbrot, B. B. 1982 *The fractal geometry of nature.* San Francisco: Freeman.
Montroll, E. W. & Shlesinger, M. F. 1984 In *Nonequilibrium phenomena II: from stochastics to hydrodynamics* (ed. J. L. Lebowitz & E. W. Montroll). Amsterdam: North Holland.
Montroll, E. W. & Weiss, G. H. 1965 *J. math. Phys.* **6**, 167.
Pfister, G. & Scher, H. 1978 *Adv. Phys.* **27**, 747.
Plonka, A., Kroh, J., Lefik, W. & Bogus, W. 1979 *J. phys. Chem.* **83**, 1807.

Pohl, R. O. 1981 In *Amorphous solids* (ed. W. A. Phillips), p. 27. Berlin: Springer.
Rammal, R., Toulouse, G. & Virasoro, M. A. 1986 *Rev. mod. Phys.* **58**, 765.
Richert, R. & Bässler, H. 1985 *Chem. Phys. Lett.* **118**, 235.
Scher, H. & Lax, M. 1973 *Phys. Rev.* B 7, 4491, 4502.
Scher, H. & Montroll, E. W. 1975 *Phys. Rev.* B 12, 2455.
Schikhof, W. H 1984 *Ultrametric calculus.* Cambridge University Press.
Schnörer, H., Domes, H., Blumen, A. & Haarer, D. 1988 *Phil. Mag. Lett.* **58**, 101.
Shlesinger, M. F. 1984 *J. statist. Phys.* **36**, 639.
Stephens, R. B. 1976 *Phys. Rev.* B 13, 852.
Tauc, J. 1984 *Semicond. Semimet.* 21 B, 299.
Toussaint, D. & Wilczek, F. 1983 *J. chem. Phys.* **78**, 2642.
Weiss, G. H. 1982–83 *Separation Sci. Technol.* **17**, 1609.
Williams, G. & Watts, D. C. 1970 *Trans. Faraday Soc.* **66**, 80.
Zumofen, G. & Blumen, A. 1982 *J. chem. Phys.* **76**, 3713.
Zumofen, G., Blumen, A. & Klafter, J. 1985 *J. chem. Phys* **82**, 3198.

Discussion

D. W. SCHAEFER *(Sandia National Laboratories, Albuquerque, U.S.A.).* It should be noted that the mean-field assumptions underlying conventional chemical kinetics are fulfilled under conditions common to most chemical kinetics experiments. As long as reaction times are long compared to diffusion times, correlation effects can be ignored. Diffusion-limited reactions are not only rare, but also difficult to study because mixing times exceed reaction times in solution. Although Professor Blumen is correct in asserting that a crossover to the diffusion limited occurs as long times in the absence of mixing, reactions are usually more than 99 % complete before crossover effects are important even in unstirred reactors.

Non-classical kinetics may be important for photochemical reactions in low-dimensional spaces. There is no evidence, however, that structural information can be extracted from anomalous reaction rates.

Reference

Schaefer, D. W., Bunker, B. C. & Wilcoxon, J. P. 1987 *Phys. Rev. Lett.* **58**, 284.

A. BLUMEN. Dr Schaefer is certainly right in pointing out that care has to be exercised in applying fractal concepts to conventional chemical kinetics, a position that we fully share (see Zumofen *et al.* 1988).

It is also very gratifying to note that the awareness for the limitations of the chemical kinetics increases, and that it is by now common knowledge that mean-field assumptions are involved. Nevertheless, I still view the positions as very theoretical (not to say academic) as long as no experimental proofs for chemical decay forms, i.e. *time-dependent* measurements over a significant dynamical range (say two orders of magnitude, at least) are provided. As long as this information is lacking, it is difficult to differentiate between beliefs and hard facts in chemical reactions.

On the other hand, a large set of photochemical decay measurements (where single-photon counting enables one to monitor around four order of magnitude in the decays) allows clear differentiations between two- and three-dimensional geometries and between different microscopic interactions. I think it is too pessimistic to rule out the possibility of extracting such information also from chemical reactions.

Here the cited work (Schaefer *et al.* 1987) and the reply to it (Even *et al.* 1987) only highlight the present difficulties, possibly also enhanced by the use of not-too-well defined materials.

References

Even, U., Rademann, K., Jortner, J., Manor, N. & Reisfeld, R. 1987 *Phys. Rev. Lett.* **58**, 285.
Schaefer, D. W., Bunker, B. C. & Wilcoxon, J. P. 1987 *Phys. Rev. Lett.* **58**, 284.
Zumofen, G., Blumen, A. & Klafter, J. 1988 In *Fractal patterns in chemistry* (ed. A. Amann *et al.*) NATO ASI-series, vol. C235, p. 1. Holland: Kluwer.

Ingram Content Group UK Ltd.
Milton Keynes UK
UKHW020346140423
420152UK00005B/250